EMERGING TECHNOLOGIES AND ETHICAL ISSUES IN ENGINEERING

Papers from a Workshop

October 14–15, 2003

NATIONAL ACADEMY OF ENGINEERING
OF THE NATIONAL ACADEMIES

THE NATIONAL ACADEMIES PRESS
Washington, D.C.
www.nap.edu

THE NATIONAL ACADEMIES PRESS 500 Fifth Street, N.W. Washington, DC 20001

NOTICE: The project that is the subject of this report was approved by the Governing Board of the National Research Council, whose members are drawn from the councils of the National Academy of Sciences, the National Academy of Engineering, and the Institute of Medicine. The members of the committee responsible for the report were chosen for their special competences and with regard for appropriate balance.

This workshop and proceedings were funded by Dr. Charles J. Pankow. Any opinions, findings, conclusions, or recommendations expressed in this publication are those of the author(s) and do not necessarily reflect the views of the National Academy of Engineering.

International Standard Book Number 0-309-09271-X (Book)
International Standard Book Number 0-309-54469-6 (PDF)

Library of Congress Catalog Card Number 2004110693

Additional copies of this report are available from the National Academies Press, 500 Fifth Street, N.W., Lockbox 285, Washington, DC 20055; (800) 624-6242 or (202) 334-3313 (in the Washington metropolitan area); Internet, http://www.nap.edu

THE NATIONAL ACADEMIES
Advisers to the Nation on Science, Engineering, and Medicine

The **National Academy of Sciences** is a private, nonprofit, self-perpetuating society of distinguished scholars engaged in scientific and engineering research, dedicated to the furtherance of science and technology and to their use for the general welfare. Upon the authority of the charter granted to it by the Congress in 1863, the Academy has a mandate that requires it to advise the federal government on scientific and technical matters. Dr. Bruce M. Alberts is president of the National Academy of Sciences.

The **National Academy of Engineering** was established in 1964, under the charter of the National Academy of Sciences, as a parallel organization of outstanding engineers. It is autonomous in its administration and in the selection of its members, sharing with the National Academy of Sciences the responsibility for advising the federal government. The National Academy of Engineering also sponsors engineering programs aimed at meeting national needs, encourages education and research, and recognizes the superior achievements of engineers. Dr. Wm. A. Wulf is president of the National Academy of Engineering.

The **Institute of Medicine** was established in 1970 by the National Academy of Sciences to secure the services of eminent members of appropriate professions in the examination of policy matters pertaining to the health of the public. The Institute acts under the responsibility given to the National Academy of Sciences by its congressional charter to be an adviser to the federal government and, upon its own initiative, to identify issues of medical care, research, and education. Dr. Harvey V. Fineberg is president of the Institute of Medicine.

The **National Research Council** was organized by the National Academy of Sciences in 1916 to associate the broad community of science and technology with the Academy's purposes of furthering knowledge and advising the federal government. Functioning in accordance with general policies determined by the Academy, the Council has become the principal operating agency of both the National Academy of Sciences and the National Academy of Engineering in providing services to the government, the public, and the scientific and engineering communities. The Council is administered jointly by both Academies and the Institute of Medicine. Dr. Bruce M. Alberts and Dr. Wm. A. Wulf are chair and vice chair, respectively, of the National Research Council.

www.national-academies.org

ORGANIZING COMMITTEE

DEBORAH G. JOHNSON (chair), University of Virginia, Charlottesville, Virginia
JOHN F. AHEARNE, Sigma Xi, The Scientific Research Society, Research Triangle Park, North Carolina
STEPHANIE J. BIRD, *Science and Engineering Ethics*, Cambridge, Massachusetts
WM. A. WULF, National Academy of Engineering, Washington, D.C.

DEDICATION

Preface

In both the popular media and the scholarly literature, scientists and engineers, policy makers, and science fiction writers and futurists have paid much attention to new and "emerging" technologies. The emerging technologies (the likely outcomes of promising new lines of research and development) include not only more information technology, nanotechnology, biotechnology, and neurotechnology, but also convergences of these diverse streams of research and development. Whether or not these new technologies develop as projected, the knowledge acquired in their pursuit is likely to have profound effects on the way humans live and think about themselves and the natural world. Separately or combined, info- , nano- , neuro- , and biotechnologies raise compelling, daunting, and unwieldy ethical and social issues. The workshop convened on October 14 and 15, 2003, by the National Academy of Engineering (NAE) was an attempt to open a discussion about the implications of emerging technologies for the future.

Engineers play a pivotal role in thinking about and bringing about future technologies. Their expertise and experience are essential to understanding the meaning and implications of new technologies. Engineers are in a unique position to comprehend, assess, and shape these technologies and to inform the public about them. But engineers are generally not experts in addressing the social and ethical implications of technology. Ethicists, humanists, and social scientists are trained to think about social meaning, social practices, and social institutions, but they are generally not equipped to understand technologies, especially new, emerging, and converging technologies. Thus, initiating a meaningful dialogue on the ethical and social issues in emerging technologies requires bringing together individuals with a wide range of perspectives, including but not limited to engineers and ethicists.

This, then, was the goal of the NAE workshop—to bring together a group of experts in different disciplines to facilitate a discussion based on an accurate understanding of current research and development and fundamental ethical concepts and approaches. The workshop was understood to be a first step, an attempt to "get ahead of the curve" by addressing the ethical and social issues raised by emerging technologies while they are still emerging. Technological development often takes place without public discussion leaving consumers and citizens to react to already developed technologies when they arrive on their doorstep. At that point, whether the new technology is well received, rejected, or greeted with a mixed reaction, it is difficult to move the technology in a new direction. A strong negative reaction may even cause the public to resist future versions of the technology—think of the public reaction to genetically modified foods and nuclear power. A strongly positive reaction may lead the public to cling to a technology that turns out to have serious, negative effects. Think of gasoline-powered automobiles, for example. To avoid these and other pitfalls, it is important that we have a public discussion about emerging technologies while they are still being developed.

NAE's mission is "to promote the technological welfare of the nation by marshaling the knowledge and insights of eminent members of the engineering profession." A workshop on the ethical issues surrounding emerging technologies promotes the realization of this mission by involving NAE members in a public discussion of the role of technology in the future of the nation. To facilitate participation by NAE members, the workshop was held immediately after an NAE annual meeting. Other participants included individuals selected from a wide range of fields. Information about the workshop was also broadly publicized, and attendance was free. Although the time between the announcement and the workshop was relatively short, more than 120 individuals attended.

The program was arranged to bring together descriptions of new technologies, the state of the art in engineering ethics, and engineering ethics education. Following the presentations, small group discussions gave participants an opportunity to think through the potentials of new technologies. On the last afternoon of the workshop, the discussion groups and a panel of the workshop organizers presented the most important ideas they had heard during the workshop. This was followed by an open session to discuss next steps.

Throughout the two days of the workshop, the discussions were intense, and passions and enthusiasm ran high. Indeed, there seemed to be few, if any, barriers to interactions among engineers, ethicists, policy makers, academics, and people from the private sector. The discussions were lively and free-flowing. Each participant seemed anxious to express an opinion about what new technologies would mean for our nation and for humanity.

The papers included in this volume appear in the order in which they were presented at the workshop. The first group introduced new technologies and the ethical issues they raise. The next set of papers focused on the state of the art in engineering ethics. The group then broke up into discussion groups guided by one of two sets of questions:

GROUP A

1. Your task is to focus on the connections between small-scale technologies, such as nano- and neurotechnologies, and engineering ethics.
2. Are there important emerging small-scale technologies that were not mentioned in the morning session that are likely to raise significant ethical issues in the future?
3. What can/should the engineering community, especially engineering organizations, such as NAE and other professional engineering groups, do to ensure that these issues are adequately addressed as these technologies are developed?

GROUP B

1. Your task is to focus on the connections between large-scale technologies, such as technologies that affect sustainability and resources, and engineering ethics.
2. Are there emerging technologies that were not mentioned in the morning session that are likely to raise significant ethical issues in the future?
3. What can/should the engineering community, especially engineering organizations, such as NAE and professional engineering groups, do to ensure that these issues are adequately addressed as these technologies are developed?

On the evening of the first day, participants were shown a film, *Incident at Morales: An Engineering Ethics Story*, which was produced by the National Institute for Engineering Ethics and the Murdough Center for Engineering Professionalism and College of Engineering of Texas Tech University. The second day began with reports from the breakout groups. These were followed by presentations on ethics in engineering education. The workshop ended with a panel presentation by four members of the workshop planning committee. Each panelist presented a summary of important ideas that had been raised during the workshop. The open discussion that followed focused on messages to take home and the next steps.

Perhaps the most striking aspect of this discussion was the easy exchange of ideas among such a diverse group. Indeed, there seemed to be a strong consensus about the importance of the issues and the value of engineers and ethicists talking to one another. One of the ideas for a next step was for more forums of this kind and more opportunities for engineers and ethicists to talk and work together.

Other ideas for next steps included: the creation of an NAE engineering ethics center; the inclusion of ethicists on NAE committees; changes in the engineering education curriculum; and making sure that resources are available on the Web.

Many of the comments related to the challenges ahead. These included: "invisibility" of many new technologies; the "messiness" of deciding when a product is good enough; the need for engineering ethics to focus more on macroethical issues, rather than microethical or individual ethical issues; the need for engineers to pay more attention to public expectations; the importance of including underrepresented groups in deliberations about the kind of world we are making; and the role of insurance companies in the products based on new technologies.

Deborah G. Johnson
Chair, Organizing Committee

Contents

ETHICS IN ENGINEERING EDUCATION

APPENDIXES

Keynote Address

WILLIAM A. WULF
National Academy of Engineering

When I looked carefully at the attendance list for this symposium, I realized that a number of you probably have only a vague idea of what the National Academy of Engineering (NAE) is. So, let me give you my elevator speech about the National Academies. Most academies of science and academies of engineering around the world share two properties. First, they are private organizations, not part of their governments. Second, they are honorific. You cannot join the Royal Society in London or the Academie des Sciences in Paris. You have to be elected by the existing membership, and that election is generally considered a high honor.

Back in 1863, in the middle of the Civil War, a group of Americans got together and created the National Academy of Sciences. They incorporated it in the District of Columbia as a not-for-profit corporation. You may remember from your high school civics class that until thirty years ago there was no city government in Washington—the federal government acted as the city government. Thus, the Academy's articles of incorporation were actually a bill passed by Congress and signed by Abraham Lincoln.

Somebody inserted about 40 words into this otherwise boilerplate corporate charter, and those words made all the difference. In modern English, they say that the academy will provide advice to the federal government on any issue of science or technology, whenever asked, and without compensation. On that basis, the academy became schizophrenic. It has two distinct personalities. On the one hand, it is an honorific organization like other academies around the world. On the other hand, it is an unbiased, absolutely authoritative advisor to our nation.

Fast forward to today. What was then one academy, the National Academy of Sciences, is now three academies—the National Academy of Sciences, the National Academy of Engineering, and the Institute of Medicine—all honorific organizations. Collectively, these three organizations manage a fourth organization, an "operating arm" called the National Research Council (NRC) that organizes most projects and studies that provide advice when the government asks for it. Together, the four organizations are now called the National Academies.

When the federal government asks us a question, we put together a committee of 10 to 20 people, literally the best people in the country on whatever the subject is, who participate pro bono. They must have no conflicts of interest, and their biases are carefully balanced. They then deliberate for anywhere from three months to three years, depending on the subject. Finally, they write a report, which I think of as a Ph.D. dissertation. It is usually 200 to 300 pages long, and the last 20 pages are citations to the literature. This fact-based, tightly reasoned report is then reviewed by a group of peers, people as eminent as the committee members themselves. The National Academies produces one of these reports every working day—about 280 last year. If you take a snapshot of the organization, there are 6,000 to 10,000 experts serving on these committees—a veritable Who's Who of people in the science and technology community, doing the best they can to serve their nation. So, that's who we are.

Now I will turn to my real topic, the subject of this gathering. We just had the annual meeting of the National Academy of Engineering, at which the president is expected to deliver an address on an issue of importance to the engineering community. For the 1999 meeting, since we were about to approach the transition to the millennium, I decided to talk about the accomplishments of engineers in the twentieth century and the challenges facing them in the twenty-first century. Preparing the first part of the lecture was easy. We had made an arrangement with the engineering professional societies to collaborate on producing a list (everybody was making lists then, if you remember). Our list was of the 20 greatest engineering achievements of the twentieth century, defined not in terms of technology "gee whiz," but in terms of impact on quality of life.

I could easily just trot out that list in my lecture, but I'm not going to repeat it here. But it is amazing! It includes electrification, automobiles, airplanes, radio and TV, agricultural mechanization, refrigeration, and on and on. The striking thing about the list is how profoundly the items on it have transformed our daily lives. If you imagine taking any one thing off of the list, you quickly realize how different life would be.

My favorite item on the list was ranked number four—and that is simply clean water. The average life span in 1900 in the United States was 46. It is now 77 plus—a difference of more than 31 years. It has been estimated that 20 of those 31 years are attributable simply to clean water and sanitation—the most prosaic engineering you can think of. In 1900, waterborne diseases were the third largest cause of death in this country. They still are in developing countries.

Almost nothing we could do for developing countries would have a bigger impact than supplying them with clean water.

Preparing the second half of the lecture, the challenges of the twenty-first century, was more difficult. Everybody agrees that the pace of technological change seems to be accelerating. Like a lot of other people, I can "guesstimate" what things will be like 10 years from now, but predicting the engineering challenges for the whole of the twenty-first century is daunting. As I looked back on the achievements in the twentieth century, I was struck by two things. First, as I have already indicated, I was in awe of how much engineering and engineers *matter*—how much they affect our daily lives. Second, I realized that the immense societal impact of engineering achievements was almost never predicted by their inventors. As Norm Augustine, a former CEO of Lockheed Martin, wrote in *The Bridge* (the quarterly publication of the NAE), "The bottom line is that the things engineers do have consequences, both positive and negative, sometimes unintended, often widespread, and occasionally irreversible."

The more I thought about that, the more I realized that there are deep moral and ethical responsibilities associated with the impact of engineering on individuals and on society. So as I searched for the second half of my speech, I began to read broadly and deeply about engineering ethics and applied ethics. In the end, I was convinced that I should pose only one challenge for the twenty-first century—engineering ethics. The quickening pace of technological innovation, the spread of nano-, bio-, and information technology, coupled with the vastly increased complexity of systems engineers are building, I now believe raise a new class of ethical questions that the engineering profession hasn't thought about. But we need to think about them, and, in fact, the need is urgent! In particular, we need to think about issues that go beyond the ethical behavior of individual engineers; we need to think about ethical behavior for the profession as a whole.

After my speech in October 1999, I also became convinced that the NAE is the one pan-disciplinary organization that has the standing and prestige in the engineering community to take on this issue—to start our fellow engineers thinking about it. With the enthusiastic backing of the NAE Council, I asked Norm Augustine to chair a committee, which Deborah Johnson, chair of this workshop steering committee, served on, to suggest how we should proceed. This meeting today is one result of the committee's report, one step in a process I hope will lead to the establishment of a permanent center on engineering ethics here at NAE.

Let me back up now and go into a little more depth. First, I don't think there is a crisis. I believe that engineers are, by and large, ethical individuals. Ethics courses are taught at many engineering schools, and there is a large literature on the subject. In addition, every engineering society has a code of ethics that usually starts with some words from the National Society of Professional Engineers code, "Hold paramount the health and welfare of the public." I think this captures very well the overall responsibility of individual engineers. These codes typically elaborate an engineer's responsibilities to clients and employers—to report dangerous or lax practices, to respect the consequences of a conflict of interests, and so on.

Beyond those codes, there are daily discussions. I can remember talking with my father and my uncle, who were both engineers, about ethical issues ranging from appropriate safety margins to undue pressure from management. I have similar vivid memories of discussions with my professors when I was in school, and with my colleagues. I remember late night debates on the subject with my friends in college.

All of that is still in place, and it's one of the reasons I'm proud to be an engineer. Individual engineers take ethics seriously. But engineering is changing in ways that raise issues that are not covered by existing codes or discussions or the textbooks I have read. These new issues are called macroethical questions (as opposed to microethical questions). "Macro" and "micro" are not intended to suggest big and important versus small and unimportant. A microethical question refers to the behavior of an individual, whereas a macroethical question refers to the responsibilities of a profession as a collective group.

The changes I want to discuss are the macroethical issues—the ones that raise questions for the profession as a whole, rather than for an individual. For engineers in the audience for whom this distinction may not be transparent, let me give you an analogy with the medical profession. The Hippocratic oath, which focuses on the behavior of individual physicians, is similar in a lot of ways to the ethical codes of professional engineering societies. But medicine is also grappling with some macroethical questions—for example, allocation. If there are not enough organs for all of the patients who need transplants, who should get them? If there is not enough medicine for all of the patients who need it, who should get it? If there are more patients than there is time for the physician to treat them, who should be treated, and on what basis?

These are not questions an individual physician decides for himself or herself. They are questions the profession must grapple with, or maybe society, guided by the profession. Once a decision is made, a physician's decision to follow that decision (or not to follow it) becomes a microethical question.

Several things have changed, and are changing, in engineering that raise macroethical questions. I'm going to talk only about the one that is closest to my professional experience—complexity. The level of complexity of the systems we are engineering today, specifically systems involving information technology, biotechnology, and increasingly nanotechnology, is simply astonishing. When systems reach a sufficiently high level of complexity, it becomes impossible to predict their behavior. It's not just hard to predict their behavior, it's *impossible* to predict their behavior. The question can't be answered by taking more things into account or thinking harder about the problem or using a new set of tools. At a certain threshold of complexity, it becomes impossible to predict all system behaviors.

Over the last several decades, mathematical theories of complexity have developed. Although these are relatively immature compared to the mathematical tools most engineers are familiar with, one thing is crystal clear. There is a level of complexity beyond which it becomes impossible to predict the behavior of systems. Unfortunately, these theories carry some undeserved baggage. For example, the term for anticipated, unpredictable behaviors is "emergent properties," a term that was first used in the 1930s in conjunction with attempts to explain why group behavior was different from individual behavior. As I understand it, these theories of group behavior are now discredited. Some postmodernists have tried to discredit science, specifically reductionist approaches to science that use similar terminology. Nevertheless, the results of these theories are solid. It is impossible, or to use the correct technical term, "intractable" to predict the behavior of sufficiently complex systems.

Let me give you an example from my own field—software. I find it fascinating that the general public tolerates a large number of errors in computer software. At any given moment, there are roughly half a million to a million bugs in the Microsoft Office suite, for example. Most people do not understand that only some of these bugs are blatant errors. Some of them are emergent properties—properties that could not be predicted. Let's talk about impossibility for a moment. Just for a touchstone, there are about 10^{100} atoms in the universe. The number of "states" in my laptop—that is, the number of patterns of zeros and ones in its primary memory—is $10^{100,000,000,000,000,000,000}$. That is an unimaginably large number! This raises an interesting problem about testing.

But first, there is something else you need to understand. Engineers will understand this better than ethicists perhaps, but physical systems have a wonderful property called "continuity." Basically, that means that for most mathematical functions that describe physical systems, if you make a small change in the input, you get a small change in the output. In other words, around a given point, continuous functions have pretty much the same value. They don't do anything radically different. The trouble with digital systems like my laptop is that they are not continuous. If you change even one bit in the memory, the meaning of what is being represented may be radically changed.

The lack of continuity has profound implications for testing. In testing physical systems, you can pick a finite number of test points that are sufficiently closely spaced, and, because of continuity, you can be reasonably certain the behavior in between those points will be similar. You cannot do that with digital systems. You have to test *every* configuration. But that is impossible; if every atom in the universe were a computer, and every computer in the universe could test 10^{100} states per second, there wouldn't be enough time, even starting from the time of the Big Bang, to test all of the states in my laptop! We have a procedure that you could follow, but there isn't enough time.

The question then becomes how to engineer software ethically when you know ahead of time that there will be behaviors you cannot predict. You cannot test for all of them, and some of them will be undesirable, possibly even disastrous.

Take another example—the U.S. Army Corps of Engineers is about to undertake an exercise to "remediate" the Everglades. We have "screwed up" the Everglades by draining them to make places where people can live, work, and shop, and the Corps is now going to "fix" them—or so they claim. But the Everglades are at least as complicated as my laptop. We don't understand the Everglades system, and we cannot predict all of the behaviors that will result from particular modifications. How can we make ethical decisions when we cannot predict what the outcomes will be? Yet doing nothing is, in fact, also doing something. We do not have the option of not doing anything and avoiding the ethical choice.

My time is just about up, so I'll have to conclude. Engineers have made tremendous contributions to our current quality of life. Certainly, we have made missteps, and certainly we need to do a lot more to bring the benefits we enjoy in the developed world to people in the developing world. I am unabashedly optimistic that we will do that, but progress is not guaranteed. We face many challenges, among them understanding what the process of engineering should be so we can engineer ethical solutions to the world's problems.

I happened on a quote from John Ladd, an emeritus professor of philosophy at Brown University, that seems apropos. "Perhaps the most mischievous side effect of ethical codes is that they tend to divert attention from the macro ethical problems of a profession to its micro ethical ones."

Thank you.

Emerging Technologies

Engineering and Ethics for an Anthropogenic Planet

BRADEN R. ALLENBY[1]
AT&T

A principal result of the Industrial Revolution, and the accompanying changes in human demographics, cultures, technology, and economic systems, has been that major natural systems are increasingly dominated by human activity. As far as we know, a planet thus impacted by a single species—the anthropogenic Earth—is a unique phenomenon. To ensure the reasonable stability of human and natural systems, which are now in many cases so integrated that the distinction between "human" and "natural" is more ideological than real, requires responsible, rational, and ethical design and management. The need for Earth systems engineering and management is apparent, but it is also apparent that the current science and technology base, institutional and governance structures, and ethical and philosophical traditions are inadequate to the task. This is not surprising because the anthropogenic Earth is unprecedented and thus requires new thinking; this is a particular challenge because tradition, ideology, and even theology combine to encourage us to turn a blind eye to what our species has wrought, with the unfortunate effect of precluding an ethical and rational response. After all, we cannot respond ethically to that which we refuse to perceive. As Heidegger (1977) cautioned:

[1]Braden R. Allenby is Environment, Health and Safety Vice President for AT&T, an adjunct professor at the University of Virginia School of Engineering and Princeton Theological Seminary, and a Batten Fellow at the University of Virginia Darden Graduate Business School. The opinions expressed herein are the author's and not necessarily those of any institution with which he is affiliated.

So long as we do not through thinking, experience what is, we can never belong to what will be. . . . the flight into tradition, out of a combination of humility and presumption, can bring about nothing in itself other than self deception and blindness in relation to the historical moment.

This paper is an attempt to explore the outlines of an ethical response and to suggest a deep connection between engineering (in the sense of understanding and designing complex systems) and the ethics appropriate to an anthropogenic world and the task of Earth systems engineering and management that lies before us.

THE ANTHROPOGENIC EARTH

For thousands of years, humans have altered the evolutionary paths of natural systems. Humans probably played a crucial role in the elimination of megafauna in Australia and North America, as well as in the disappearance of prey species, such as the moas of New Zealand (Jablonski, 1991; Perkins, 2003). Ice deposits in Greenland show spikes in copper concentrations reflecting the production of copper during the Sung Dynasty in China (ca. 1000 B.C.); lake sediments in Sweden similarly reflect the production of lead in ancient Athens, Rome, and medieval Europe (Hong et al., 1996; Renberg et al., 1994). Anthropogenic buildup of carbon dioxide in the atmosphere had been going on for millennia with the deforestation of Eurasia and Africa, although concentrations were clearly accelerated with the Industrial Revolution and subsequent reliance on fossil fuels (Grubler, 1998; Jager and Barry, 1990). The long evolution of agriculture is also a history of increasing anthropogenic impacts, both intended and unintended, on natural systems (Redman, 1999).

The Industrial Revolution cemented the rise of the anthropogenic Earth. The enormous expansion of human activity and influence on natural systems resulting from the Industrial Revolution, and the advent of a global, highly technological, market-oriented world culture, are only hinted at in the data. In terms of global gross domestic product (GDP), if 1500 A.D. is indexed at 100, by 1992 world GDP had risen to 11,664—more than a hundred-fold increase. If 1900 is indexed as 100, total energy use in 1800 was only 21—but in 1990 it was 1,580. In 1700, total global freshwater withdrawals are estimated to have been around 110 cubic kilometers; by 2000 they were estimated to be 5,190 cubic kilometers (all figures from McNeill, 2000). Technology evolved from modest beginnings in textile production through steam power and iron production into the large-scale mass production of consumer goods, such as automobiles, and it continues to expand. The advent of major new technologies, especially nanotechnology, biotechnology, and information and communications technology, and continued advances in the cognitive sciences will extend human design capabilities in the coming decades into new realms—the very small, life itself, and an all-encompassing cybersphere.

The effects of these technologies are apparent. Humans have already begun a dialogue with the climate cycle, and implicitly the carbon cycle, a necessary dialogue in light of the effect of human activity on the dynamics of these systems, which requires conscious design and management. Other critical cycles, such as the hydrologic cycle and nitrogen cycle, are similarly affected, although so far the response to these effects has been less organized. With genetic engineering and proteomics, the biosphere at all scales is increasingly becoming a subject of human design (Science, 1999a,b). From a purely physical perspective, the human transport of soil and rock is of the same magnitude as transport by natural forces, such as wind and water erosion, sediment transport, glaciers, and volcanoes (McNeill, 2000). As Gallagher and Carpenter (1997) remark in their introduction to a special issue of *Science* on human-dominated ecosystems, the idea of pristine ecosystems untouched by human activity "is collapsing in the wake of scientists' realization that there are *no places left on Earth that don't fall under humanity's shadow*" emphasis added). Palumbi (2001) similarly comments that: "Human ecological impact has enormous evolutionary consequences . . . and can greatly accelerate evolutionary change in the species around us . . . [T]echnological impact has increased so markedly over the past few decades that humans may be the world's dominant evolutionary force."

Awareness that the Earth is indeed anthropogenic—or that we are now in the "anthropocene" (Nature, 2003)—is not new[2] or a sudden discontinuity. It is the culmination of 2,500 years of human cultural and technological evolution.[3] Indeed, one can identify the stirrings of an institutional response in developments such as "adaptive management," a nascent management approach based on attempts at the comprehensive management of regional resource complexes, such as the Baltic Sea, the Everglades, Canadian forests, and global fisheries. This evolving practice is defined in a leading text (Gunderson et al., 1995) as providing "ways for active adaptation and learning in dealing with uncertainty in the management of complex regional ecosystems" (see also Berkes and Folke, 1998).

Others have begun to establish operational principles for Earth systems engineering and management (Allenby, 2000/2001, 2002). Some of these, like the observation that even "natural" systems like the Everglades are now products of human design and choice, and will continue to be so for the foreseeable future,

---

[2]Thus, for example, W.L. Thomas 1956 classic, *Man's Role in Changing the Face of the Earth*, and the 1989 special issue of *Scientific American* entitled "Managing Planet Earth."

[3]As Barrett (1979) comments, "A great chapter in human history—twenty-five hundred years long, from the beginnings of rational thought among the Greeks to the present—has come to an end. . . . [a situation that] calls us toward some other dimension of thinking of which we can catch now and then perhaps only glimmers." In fact, our challenge is precisely to learn that dimension so that we may engineer and manage what we have already brought about—and do so rationally, responsibly, and ethically.

have significant implications for engineering practice. In such cases, the model for engineering and management is not focused on creating a defined end point (an engineered artifact, for example) but is an ongoing dialogue with the systems involved. This dialogue includes not only scientific and engineering dimensions, but also policy and cultural dimensions; it is a dialectic that is not familiar to engineers, policy makers, or the public. At this point, no one knows how to do it or is even able to accept it. For example, regardless of the particular actions we may take at any point in time, we will be in a constant dialogue for the foreseeable future with the climate system (and, thus, the carbon and nitrogen cycles, among others), a dialogue that involves design, responsibility, and ethics, and thus is a case study in Earth systems engineering and management.

But the operational aspects of the dialogue, new and critical as they may be, are not the focus of this paper. We must now ask an even deeper question—what ethical structure we can develop to support our responses to the anthropogenic planet, this terraformed Earth. I will start by considering the currently popular concept of "sustainability" and then move beyond that to suggest a fundamental coupling of the engineering worldview and the ethical foundations necessary for Earth systems engineering and management. My remarks are both preliminary and schematic and may well be incomplete or conceptually flawed. Nevertheless, they may initiate a dialogue that must become part of our skills as engineers. The anthropogenic world is not a hypothetical that can wait on academic musings. It lies before us even now.

SUSTAINABLE DEVELOPMENT AND CULTURAL CONSTRUCTS

Let us begin with a simple question. What is "sustainable development," or, more broadly, what is "sustainability"? The specifics can be supplied. The phrase "sustainable development" was introduced by the World Commission on Environment and Development, also known as the Brundtland Commission, in 1987 in *Our Common Future*; sustainable development was defined as development that "meets the needs of the present without compromising the ability of future generations to meet their own needs" (WCED, 1987). As initially stated, sustainable development was understood to require a more equitable distribution of resources and limits on consumption.[4]

But the formulation was somewhat vague, and an outpouring of explanations, new meanings, commentaries, definitions and redefinitions, and expositions ensued. There were two results. First, it became fashionable to use the word

---

[4] "Meeting essential needs requires not only a new era of economic growth for nations in which the majority are poor, but an assurance that those poor get their fair share of the resources required to sustain that growth" (WCED, 1987, p. 8). Regarding the wealthier societies, the commission noted, "Sustainable global development requires that those who are more affluent adopt life-styles within the planet's ecological means . . . " (WCED, 1987, p. 9).

"sustainable" to describe virtually any entity. Thus, one now hears about sustainable cities, sustainable firms, sustainable practices, sustainable artifacts of all kinds, sustainable campuses, sustainability science, and so on. Second, the proliferation of meanings meant that the idea of sustainability became so ambiguous as to be almost meaningless; the difference between a "sustainable X" and a "just plain X" was not at all clear. At best, the adjective now indicates a generally supportive attitude towards environmentalism, and, most of the time, a mild impulse toward redistribution of wealth.

Sustainable development is thus a classic example of a cultural construct, a concept contingent on a particular time, place, and culture reflecting a particular set of values. Cultural constructs usually have a number of purposes (Hacking, 1999). A major motivation in this case was that environmentalism (the determination to reduce the environmental impacts of economic growth) was increasingly feared to be in conflict with economic growth, especially in developing countries. The cultural construct of sustainable development, on the contrary, implies that they are not necessarily in conflict but can be integrated, at least linguistically; however, a cultural construct does not necessarily mean that the underlying conflicts in the external environment have been resolved. The cultural construct also provides a vehicle by which the underlying political and ideological discourses can be advanced—in this case, social democratic impulses toward egalitarianism as opposed to libertarianism (that is, equality of outcome, as opposed to equality of opportunity) and environmentalism.

Cultural constructs are not inherently undesirable; indeed, they are necessary, for they provide a way to make an otherwise intractably complex world intelligible. And they are pervasive in environmental discourse (as in all discourse). Thus, it is not surprising that terms that are considered self-evidently "real," such as "nature" or "wilderness," are in fact highly contingent cultural constructs. As C. Merchant (1995) observed in her essay "Rethinking Eden," "Nature, wilderness and civilization are socially constructed concepts that change over time and serve as stage settings in the progressive narrative." And W. Cronon, in his classic book *Uncommon Ground* (1995), identifies at least 10 separate meanings packed into "nature," the most powerful perhaps being the theological evolution of nature into something sacred:

> This habit of appealing to nature for moral authority is in large measure a product of the European Enlightenment. By no means all people in history have sought to ground their beliefs in this particular way. . . . the fact that so many now cite Nature instead [of God] (implicitly capitalizing it as they once may have capitalized God) suggests the extent to which nature has become a secular deity in this post-romantic age (p. 36).[5]

---

[5]For a discussion of how the sacred was shifted from God to Nature by the Romantics during the European Enlightenment, partly in an attempt to defend medieval Christian theology in light of scientific advances, see Abrams (1971).

Perhaps the most interesting shift in the meaning of nature, however, is in how its opposite is defined. For hundreds of years, the opposite of natural was supernatural (e.g., ghosts, ghouls, etc.). Today, however, the opposite of natural is often defined as human; a natural food has minimal manufactured inputs, and a natural fiber is made of material not synthesized by humans. The implications of this change for engineering are obvious; for some, at least, human creativity, including, and perhaps especially, engineering, has been placed firmly in the camp of the unnatural.

Environmental discourse offers many other interesting examples of similar shifts. A quasipermanent area of waterlogged land has evolved from a swamp (pestilent, economically useless, and therefore evil) to wetlands (useful, biologically productive, and therefore good); and a tropical forested area has evolved from a jungle (again pestilent, dangerous, a place of death, and therefore evil) to a rain forest (a place of life and gentle mist, Edenic, and without the human stain) (Allenby, 2002; Cronon, 1995; Merchant, 1995). Two hundred years ago, wilderness was considered evil, satanic, the result of the biblical Fall; the first European settlers in the Americas saw their religious obligation as the conversion of the surrounding forests into farmland and gardens, the creation of a New Jerusalem out of chaos and night. As John Quincy Adams said in his 1846 appeal to Americans to settle Oregon, the mission was "to make the wilderness blossom as the rose, to establish laws, to increase, multiply, and subdue the earth, which we are commanded to do by the first behest of the God Almighty" (Merchant, 1995; Sagoff, 1996). Today, preserving wilderness is a principle public policy goal.

Just because cultural constructs are ubiquitous and necessary does not mean that they are benign. Consider, for example, the logical implications of redefining natural as nonhuman and wilderness as an Eden devoid of people. The obvious corollary is that further changes of nature and wilderness are satanic and must be halted immediately (in extreme cases, this provides a psychological framework for people like the Unibomber and members of the Earth Defense League who feel morally justified in attacking engineered artifacts and the people who engineer them).

Cultural constructs, then, become powerful ideological and ethical screens that identify those who are "good" (i.e., those who accept the culture embedded in the construct) and those who are "bad."[6] They can also stifle debate. On one

---

[6]This can be seen clearly in some biocentricist writings. For example, Singer (2001) writes, "if we do not change our dietary habits [to become complete vegetarians], how can we censure those slaveholders who would not change their own way of being?", thus equating a practice, slavery, that most would regard as evil, with anyone who eats any animal. Similarly from deep ecology (Berry, 2001): "a deep cultural pathology has developed in Western society . . . a savage plundering of the entire earth. . . . This plundering is being perpetrated mainly by the great industrial establishments that have dominated the entire planetary process for the past one hundred years. . . . Opposed to the

hand, it is hard to find people who declare they are against wilderness, nature, sustainability, or similar concepts; on the other hand, by accepting those terms they must accept the cultural structure embedded in them. In this respect, the language of environmentalism is like the language of Marxism. By speaking it, one is forced to accept its ethical and cultural content and its ideas about the world.[7]

Another critical aspect of cultural constructs—that they change over time— is particularly applicable to engineering at the scale of Earth systems. In the short run, cultural constructs are indeed fixed; the idea of wilderness in the United States has changed little over the past decade. But in the long run, they are fluid. The time scale of traditional engineering (e.g., designing a toaster or an automobile) is well within the time cycle of change of cultural constructs—that is, the period during which a construct remains stable. But Earth systems engineering and management (e.g., designing and supporting the continued evolution of the Florida Everglades or engineering the carbon cycle to stabilize climate variation within desired limits) is a systems function (a dialogue between human and natural systems) that extends beyond the time cycle of many cultural constructs. Thus, while we can assume a fixed cultural context for traditional design projects, we cannot assume this for Earth systems engineering and management.

Indeed, Earth systems engineering self-referentially creates its own context. This simple observation has serious consequences; instead of a stable intellectual framework, we have a self-referential, self-organizing structure that operates not only at the scientific and engineering levels, but also at the ethical and cultural levels. For traditional projects, ethical systems are implicit and in context; for the latter, ethical systems are a part of the design (and may in practice be an emergent characteristic of the design, becoming apparent only as the dialogue between designer and complex system evolves).

The ethical implications of this vastly more complicated design challenge are yet to be understood. Indeed, in many cases the conditions that require these long-term designs have yet to be universally perceived. But it is possible to begin the process by identifying the most salient characteristic of the anthropo-

---

industrial establishment is the ecological movement which seeks to create a more viable context for the human. . . ." This last passage clearly differentiates between the evil (industry, including modern science and technology) and the good (ecowarriors) and suggests a conflict between engineering and ecology through the apocalyptic structure built into the language of environmentalism.

[7]Environmentalism, like Marxism, illustrates that all languages are contingent and related to power structures (Lyotard, 1979; Rorty, 1989). Lyotard (1979) in fact speaks of the "terroristic" power of languages, in that they can silence those who have interests or values different from those embedded in a particular dominant language. Although this formulation is somewhat dramatic, given the activities of Hitler, Stalin, Mao, and Pol Pot, among others, dominant languages can stifle debate. This explains the concerns expressed by some in developing countries about attempts by Western environmentalists to impose their views and values.

genic world—complexity—and building on it. The question is not whether we wish to live in an anthropogenic world; we have already created it, and we already do. The question is whether we want to live in it ethically, responsibly, and rationally.

DESIGN, COMPLEXITY, AND POSTMODERNISM

The single most overwhelming reality of the anthropogenic world is its complexity, the static complexity of economies, cultures, and natural cycles and systems and the dynamic complexity of their internal and external unfolding as networked systems over time. In addition, complexity is introduced by the contingent and reflexive characteristics of human systems, which reflect the choices and intentions of human agents and interactions through and across other systems and networks. The sheer unintelligibility of complex human/natural systems based on our current individual and institutional perceptual and conceptual frameworks is apparent to anyone who has attempted to work rationally on complex systems, such as the Everglades, or in natural-resource regimes, such as fisheries or forests (Michael, 1995):

> Persons and organizations view information from their personal and peer-shared myths and boundaries. More information provides an ever-larger pool out of which interested parties can fish differing positions on the history of what has led to current circumstances, on what is now happening, on what needs to be done, and on what the consequences will be. And more information often stimulates the creation of more options, resulting in the creation of still more information . . .

> Indeed, in our current world situation, opening oneself or one's group to a larger 'database' reveals the terrifying prospect that the world is now so complex that no one really understands its dynamics and that even rational efforts tend to be washed out or misdirected by processes not understood and consequences not anticipated. Of course, as suggested earlier, those intent on pursuing their interests seldom can risk sociocultural ostracism by acknowledging this to others, and usually not even to themselves.

Similarly, Senge (1990) also comments on the inability of individuals in industry to comprehend their environment:

> . . . we are being overwhelmed by complexity. Perhaps for the first time in history, humankind has the capacity to create far more information than anyone can absorb, to foster far greater interdependency than anyone can manage, and to accelerate change far faster than anyone's ability to keep pace. Certainly the scale of complexity is without precedent.

Let us be clear about one point. The complexity of design in an Earth systems and management context—be it the Everglades; the climate cycle; the hydrologic, carbon, or nitrogen cycle; mega-urban systems; or the infosphere—

is not just the result of technical aspects of those systems. Climate change negotiations are not about technique. Rather, the complexity arises in large part because, once we get to this stage, we cannot avoid dealing with C.P. Snow's "two cultures"—(1) the scientific and technical culture and (2) the literary and humanist culture (Snow, 1959). The anthropogenic Earth is characterized by systems that integrate the profoundly human—economic institutions, information systems, cultures, governance systems—with the physical, biological, and chemical systems we call natural (Figure 1). It is becoming increasingly apparent that natural systems cannot be understood without knowing the human history that led to their current state; thus, for example, it is futile to try to understand the ecology of an island without understanding the human transportation systems and migration patterns that have affected it, just as it is futile to try to understand the biology of the Everglades without understanding the politics and money of the sugar industry and the demographics and settlement patterns of Florida. When we negotiate about climate, we are simultaneously negotiating about the structure we desire for the carbon cycle and about the future paths of human economic and cultural development that we will allow, and not allow. And when the deep greens insist that the United States curb its carbon emissions directly, rather than through reductions in other countries' emissions, they are trying to socially engineer U.S. consumers, and not just reduce global climate-change forcing.

The most important implication of human/natural unitary systems is that human complexity has been imported into the dynamics of fundamental natural systems. Natural systems are complex, but human systems are even more complex, an important distinction in light of recent literature that draws implicitly on the analogy between natural and ecological systems and human systems. The analogy can be useful, as the development of the field of industrial ecology has shown (Allenby, 1999; Graedel and Allenby, 2002; Socolow et al., 1994). Indeed, human and natural systems are similar in that they are both technically complex and that the lessons learned from natural systems can indeed inform our understanding of human systems in some ways. But an analogy can only take us so far. Failure to recognize the profound differences between natural and human systems can lead to superficial reasoning or even nonsense (take, for example, the burgeoning literature suggesting that global capitalism or transnational corporations can be restructured to resemble gardens).

Human systems have a different, and higher, level of complexity than natural systems. Human systems and human history are strongly affected by unpredictable contingencies, partly because we have (bounded) free will, which makes humans relatively autonomous moral agents (Hacking, 1999; Harvey, 1996; Landes, 1998; Scott, 1998). Moreover, human systems are characterized by reflexivity. A natural system, such as a salt marsh, is not changed by what a scientist learns about it, but human systems are, because knowledge is internalized as it is developed; thus human systems change continually in an accelerating

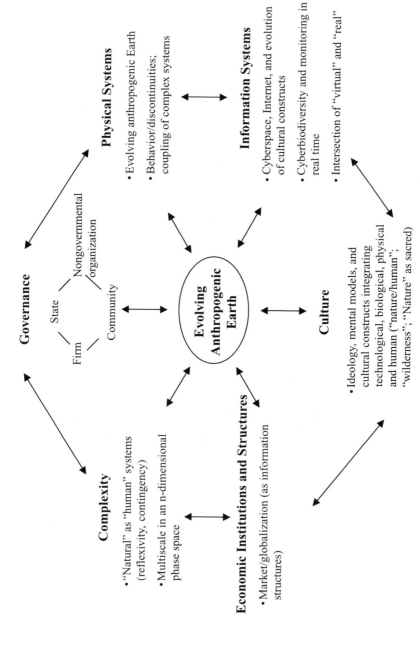

FIGURE 1 The anthropogenic Earth.

process of reflexive growth (Castells, 2000; Giddens, 1984). The evolutionary processes of culture, technology, and social knowledge are uniquely human projects, with their own dynamics and time frames; they have no parallel in traditional natural systems (Grubler, 1998; Heidegger, 1977; Noble, 1998). Thus, understanding the anthropogenic world requires not just that we understand the scientific and technological domains, but that we also understand the social science domains—culture, religion, politics, economics, and institutional dynamics.

Postmodernism, which emphasizes pastiche and the multiplicity of discourses, can be understood, in part, as a reflection into philosophy of increased complexity of human systems in the twentieth century. Every dimension of a human system is complex, including the intuitive dimension of human experience of community. As Mitchell (2000) notes, complexity has increased significantly over the past century and is accelerating as a result of the Information Revolution:

> If you live to a good age, you have maybe half a million waking hours. If your world of interaction is at a village scale, each member of it gets on average a couple of thousand hours of your time. At an automobile scale, it is down to two hours each. And at a global computer network scale, it is reduced to less than ten seconds. Clearly, then, attention becomes a scarce resource, and intervening attention management mechanisms are essential if we are not to be overwhelmed by the sheer scale at which electronically mediated global society is beginning to operate.

Not only is the world more complex, then, by orders of magnitude, but the means of perceiving and thinking about it can no longer be internal to a single human being. Individual cognition is a function of technology and social networks as much as it is of biology: "individuals position themselves less as members of discrete, well-bounded civic formations and more as intersection points of multiple, spatially diffuse, categorical communities" (Mitchell, 2000). To put it another way, postmodernism may be seen as the recognition that cognitive systems have, in a multicultural world, changed not just in degree but also in kind. Their complexity defies traditional philosophic explanation.[8]

Unfortunately, the response to this realization by many postmodernists is to abdicate responsibility and retreat into absolute relativism ("there are no privileged discourses"). This is not only unnecessary, but it is so contrary to most people's sense of reality that postmodernist discourse is confined to the intelli-

---

[8]The expansion of cognition beyond the individual obviously has many important implications, and even a cursory investigation would take us well beyond the scope of this paper. A more detailed discussion can be found in Rowlands (1999) and Allenby (2002), where the idea of "integrative cognitivism" is introduced.

gentsia; it has become an amusement, not a philosophy.[9] Indeed, this extreme relativism, and the concomitant denigration of science and engineering, frequently fuels animosity between engineers and their postmodernist critics. With a more sophisticated sense of networked complex systems, the reason for the antagonism disappears (it may well continue as a result of cultural norms and ideological posturing, but that is a separate issue).

Postmodernists tend to make a simple mistake. Because they are very sensitive to global unpredictability and contingency, they assume these properties also predominate at lesser scales. In fact, the anthropogenic world is characterized by many ordered structures that are local in time or space (usually both), even if unpredictable chaos seems to be the order of the day at greater scales.

At this point, the intuition of systems function that underlies much of engineering intersects with ethics and responsibility in the anthropogenic world. The complexity of the modern world does not mean complete disorder and thus does not imply absolute relativism. The world can be thought of as complicated, coupled, evolving systems of networks, reacting to both internal and external changes in a number of state spaces. (The most obvious dimensions of these spaces are time and space, but because the world is anthropogenic, we must add new dimensions, including, but not limited to, culture, information, and, perhaps, technology, economics, and institutions.)[10] These networks form a shifting pattern, which inevitably includes patterns of local order amidst the global disorder, and vice versa. Some structures (religions, for example) may last for thousands of years; others may be lost in seconds, minutes, or months.

Thus, ethical structures need not claim to be foundationally valid for all time and space, even though they are absolute within a particular local order, as long

---

[9]Actually, few postmodernists go to the extreme of absolute relativism, at least in their own ethical stances. It is surprising how often individual postmodernists find enough structure in the world to validate their own particular positions, even as they deride those of others. Science and technology, in particular, are a favorite target of postmodernists, in part because they dominate discourse in the globalized, Eurocentric culture; thus they must be negated if other, more literary, discourses are to become ascendant.

[10] One example might be the "actor network" that some students of technology use to describe the process of technological evolution (Callon, 1997):

> The actor network is reducible neither to an actor alone nor to a network. . . [I]t is composed of a series of heterogeneous elements, animate and inanimate, that have been linked to one another for a certain period of time. . . [T]he entities it is composed of, whether natural or social, could at any moment redefine their identity and mutual relationships in some new way and bring new elements into the network. An actor network is simultaneously an actor whose activity is networking heterogeneous elements and a network that is able to redefine and transform what it is made of.

As Allenby (2002) points out, an actor network combines the ideas of intentionality (human design, implying ethical responsibility) and systems dynamics, thus creating constraints and opportunities. In other words, the ability to exercise intentionality becomes a function of system state.

as that order persists. Consider the example of a toaster. When we say that designing a toaster is relatively simple, we mean that artifact design occurs within a pattern of local order that has established ethical norms. When we talk about designing the Florida Everglades or the climate cycle, however, we are talking about time frames that extend well beyond the boundaries of the locally stable system. We must then accept that contingent ethics and values, as well as design objectives and constraints, are part of the engineering challenge.

It is precisely the failure to recognize the profound difference between design in local order and design beyond the boundaries of local order that has caused so much difficulty in the climate change negotiations. In one case (local order), ethics are established and usually personal; in the other case (beyond local space and time), ethics are contingent, probably multicultural, perhaps mutually exclusive, and not usually exercised at the personal level. Just as a systems engineer must include as part of his or her assessment the values of different stakeholders, at a much more profound level, the Earth systems engineer must move beyond personal beliefs to appreciate and, indeed, respect the values of many systems and cultures involved in the complex, evolving system.

This requires a new and complex ethical structure, personal on one level, institutional, inclusive, and nonjudgmental on other levels. The differentiator is whether the engineering task lies within an area of local order or extends beyond local boundaries. Consider the examples in Table 1. An electrical engineer designs an Internet protocol router that becomes part of the Internet. If the router malfunctions as a result of sloppy design, one might make ethical judgments about the designer. If, however, the Internet as a system has an unanticipated effect—say, to make adolescent males less functional socially because they spend all their time playing video games—one would not be inclined to blame the engineer who designed the router. And one would certainly not put the ethical responsibility for a world increasingly defined by the infosphere as an overlay on other Earth systems (from the economy to the carbon cycle and hydrologic cycle) on the engineer who designed the router. Similarly, the shipbuilder, unknown to history, who first designed the Portuguese caravel and thus enabled oceanic travel, could be blamed if the ship sank because of poor design, but he could not be blamed for the ecological effects of the global oceanic transport system that evolved or for the eventual globalization of the Eurocentric, Christocentric culture.

And yet, individual designs now become components of (frequently self-organizing) complex networks, such as the Internet. Thus, humans are designing Earth systems of all kinds, from cultural, economic, and demographic systems to natural cycles and systems. Because these systems extend beyond the patterns of local order, yet are increasingly the products of human design taken as a whole, those responsible for the designs must also take responsibility for the results. This leads to the greatest ethical challenge of the anthropocene, for the anthropogenic Earth is characterized by human ethical and cultural systems that are increasingly becoming reified in natural systems (Figure 2).

TABLE 1 Evolution of Engineering Ethics in an Anthropogenic World

	Technology	Explicit (artifact) design	Implicit (infrastructure/ system) design	Teleology
System scale	Internet	Router, personal computer	Internet as physical and information/ cultural system	Infosphere as network for Earth systems engineering and management (the networked world)
	Sailing ship	Portuguese caravel	Global transport/ migration/ colonization system	Eurocentric globalization
	Biotechnology	Genetically engineered, salt-resistant tomato plant	Optimized biomass productivity	Life at all scales as human design
Ethical responsibility of engineer/ designer	Current	Yes, often embedded in law (e.g., product liability)	No, system effects often not knowable with current state of the art	Implicit and usually unconscious
	Earth systems engineering and management	Yes, personal	Yes, probably exercised mainly through institutions (private, public, and professional), and bounded by uncertainty, sys- tem dynamics, and state of the art	Explicit and part of education, design process, and client/ stakeholder dialogue

For example, the biological structure of the world as it now exists profoundly reflects the Christian, Eurocentric culture that has migrated and colonized the world in the centuries since the Enlightenment and the Industrial Revolution. The structure of the Everglades reflects the ethical and cultural capitalist system that has prevailed in the United States for 200 years. The

23

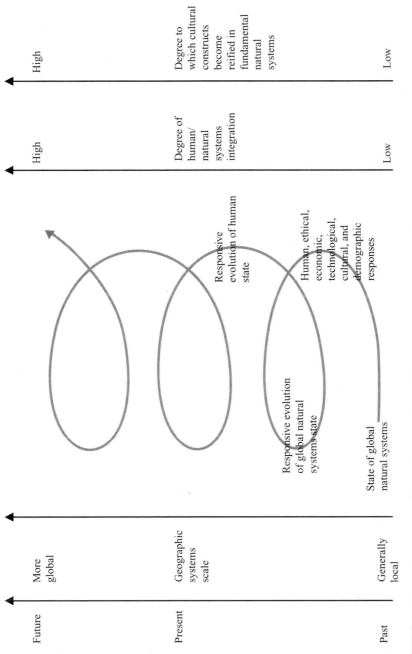

FIGURE 2 Schematic dynamics of the anthropogenic world.

Internet reflects the capitalist ideologies, market structures, and mass information production and consumption patterns of a globalized, high-technology culture.

Three trends have characterized this process: (1) human and natural systems have become increasingly integrated, to the point that, in many cases, there is no meaningful difference between them; (2) as the human capability to manipulate the external environment has increased, cultural constructs have become increasingly reified in fundamental natural systems; and (3) the physical scale of natural systems has expanded from the local to the global. In the future, human and natural systems are likely to become even more tightly integrated, and human concepts and teleologies even more reified in fundamental natural systems. A century from now, the climate system will reflect whatever values and ethics we have brought to the management of that system. Thus, "natural history" becomes human history, and ethics becomes not just a desirable adjunct to engineering, but a core competency.

In short, the ethics called for by the anthropogenic Earth are indeed different, and more complex, than "traditional" ethics, precisely because we are now aware of our collective responsibility for emergent behaviors at higher levels of the system than the level of the designed artifacts. To put it bluntly, those who engineer parts of systems are responsible in some way, both ethically and rationally, for the system as a whole.

This responsibility poses several immediate problems. First, given our current state of knowledge, the behavior of complex systems is often not predictable, or even knowable. The Internet, for example, despite its obvious human provenance, is also a self-organizing system; we don't even have a good map of it at this point (Barabasi, 2002). We know even less about the eventual cultural, demographic, and environmental effects of the Internet. It might, for example, accelerate the flow of information through human cultures, thereby accelerating the pace of changes in the meaning of cultural constructs, thus reducing the pockets of local order that give the illusion of stability most of us take for granted. And what effect might that have on environmental consciousness and cultural attitudes towards "less human" systems? Might the virtual become so powerful that it displaces the concern with the real, and might we then be content to play with virtual pandas while the real ones become extinct? More likely, the genetic information of pandas might be explicated, stored, and owned, so they can be regenerated by the owner of the information.

An ethical principle that has been enshrined in many criminal laws is that one can only be held responsible for what one knows or can reasonably foresee (Kane, 1998). Most of us balk at the idea of holding the router designer responsible for the unforeseeable dynamics of the Internet or the fifteenth-century shipbuilder responsible for European colonialism or the rise of high technology or globalized economic markets. Given the anthropogenic Earth, the traditional concept of ethical responsibility is necessary but no longer sufficient. The underlying ethical structure of engineering must be expanded so the ethical dimensions of engineered systems at the Earth-system scale can be perceived, understood, and addressed.

The difficulty of this challenge cannot be overemphasized. Even at the most basic level (the level of the individual), human psychology predisposes people to think in terms of simple systems conceptualized in terms of relatively few variables with interrelationships that can be easily and almost completely understood and that are displayed over a short period of time. In short, we are psychologically attuned to operating within the bounds of locally ordered networks. But many aspects of complex systems are difficult and counterintuitive and can only be illustrated by the behavior of properly constructed quantitative models. As Michael (1995) puts it based on his experiences with adaptive management of natural resource systems:

> *Our conventional ways of thinking and speaking about language and social reality are inadequate for coping with our current circumstances....* Our semantic baggage from past experiences is not matched to a reality of systemic interactions, circular feedback processes, nonlinearity, or multiple causation and outcomes. Implicitly, our conventional language relates us to a world of linear relationships, simple cause and effect, and separate circumstances, be they events, causes, or effects. But that is not the world we live in.

This is an extremely important point, for, as modern philosophers have pointed out, if something cannot be captured in language, it cannot be perceived and cannot be a part of our reality (Rorty, 1989).

The problem is not just that we do not understand the very complex anthropogenic world that integrates the reflexivity and contingency of human systems (economies, political systems, history, culture, religions) into natural systems or that we do not even understand human systems very well. The problem is more profound. Over the past 2,500 years, we have created an anthropogenic world that has extended the implications of our designs and engineering decisions beyond our capacity to predict, to choose, or even to perceive, their outcomes. Thus we have created a moral gap between what we actually do and what we take responsibility for.

ETHICS FOR AN ANTHROPOGENIC WORLD

It is doubtful that the burden of extending engineering ethics can be assumed solely by the engineer(s) directly involved. For one thing, the complexities of these systems necessarily engage stakeholder groups with different worldviews and disciplinary backgrounds. Each discipline not only has its own approach to ethics, but, perhaps more fundamentally, has its own ontology, or set of assumptions about the fundamental nature of being. Scientists and engineers, for example, tend to believe strongly in the external, physical world, whereas many sociologists, cultural anthropologists, and literary critics may believe equally strongly that reality is constructed by the individual and her or his society. The truly difficult challenge for Earth systems engineers is to understand that these mutually exclusive ontologies are both correct. In short, the complexity of the anthropogenic world cannot be captured in a single belief system or ontology. Different ontologies may be appropriate for different locally ordered networks, and issues that cut across patterns of local order may thus appropriately engage a number of different ontologies. We will need teamwork and institutions that approach ethical issues in a multicultural and multidimensional way. The need to rise above individually valid ontologies carries with it an implication that militates against an individual being charged with professional and personal ethical responsibility, but also with direct ethical responsibility for (uncertain and unpredictable) system performance.

Our most obvious pressing need is to begin to develop not just personal, but also institutional capabilities to address ethical issues arising from the emergent behavior of global natural/human systems. Professional engineering organizations, such as the Institute of Electrical and Electronic Engineers (IEEE), the Society of Automotive Engineers (SAE), and the American Society of Mechanical Engineers (ASME), should support multidisciplinary capabilities to identify, study, and make recommendations regarding emerging ethical issues, both specifically (e.g., the environmental and social implications of the shift from mail-based to e-mail-based consumer systems) and generally (e.g., ethical issues arising from the evolution of a wired world dominated by human information and communications technology). Existing efforts, such as bioethics institutions and panels associated with advances in biotechnology and medical technology, should be continued and indeed encouraged, and may serve as models. But they are insufficient. The National Academy of Engineering, as the thought leader for the profession both in the United States and around the world, should take the lead in this effort.

The new, expanded ethical approach must become part of modern engineering. To this end, the National Science Foundation should fund a network of academic institutions for researching and teaching ethics and engineering in an anthropogenic world. The goal should be not only to develop and publish curricular material for engineering and related fields, but also to establish a global, networked community of scholars from many disciplines. As our understanding improves, appropriate material can be incorporated into engineering education. In many cases, material may simply be added to the traditional curriculum, which will continue to educate electrical, mechanical, and civil engineers. But there will also undoubtedly be new programs focused on educating Earth systems engineers and managers.

Individual engineers will continue to be ethically responsible for their particular designs and activities. But we must also assume greater responsibility as a community. Individual ethical responsibility should be understood to include helping to create, and contribute to, professional community ethical responses at the level of systemic behavior. We now inhabit a terraformed planet that is shaped by and displays our designs, choices, and cultures at all scales, from artifacts to great natural cycles. We have an obligation to ourselves, to our profession, and to the future to create the knowledge and wisdom that will make this Earth, and the designs that define it and knit it together, the highest expression of our responsibility, our rationality, and our ethics.

REFERENCES

Abrams, M.H. 1971. Natural Supernaturalism: Tradition and Revolution in Romantic Literature. New York: W.W. Norton and Company.

Allenby, B.R. 1999. Industrial Ecology: Policy Framework and Implementation. Upper Saddle River, N.J.: Prentice-Hall.

Allenby, B.R. 2000/2001. Earth systems engineering and management. Technology and Society 19(4): 10–24.

Allenby, B.R. 2002. Observations on the philosophic implications of Earth systems engineering and management. Batten Institute Working Paper. Charlottesville, Va.: Batten Institute, Darden Graduate School of Business, University of Virginia.

Barabasi, A. 2002. Linked: The New Science of Networks. Cambridge, Mass.: Perseus Publishing.

Barrett, W. 1979. The Illusion of Technique. Garden City, N.Y.: Anchor Books.

Berkes, F., and C. Folke, eds. 1998. Linking Social and Ecological Systems: Management Practices and Social Mechanisms for Building Resilience. Cambridge, U.K.: Cambridge University Press.

Berry, T. 2001. The Viable Human. Pp. 175–184 in Environmental Philosophy: From Animal Rights to Radical Ecology, 3rd ed., edited by M.E. Zimmerman, J.B. Callicott, G. Sessions, K.J. Warren, and J. Clark. Upper Saddle River, N.J.: Prentice-Hall.

Callon, M. 1997. Society in the Making: The Study of Technology as a Tool for Sociological Analysis. Pp. 83–86 in The Social Construction of Technological Systems, edited by W.E. Bijker, T.P. Hughes, and T. Pinch. Cambridge, Mass.: MIT Press.

Castells, M. 2000. The Rise of the Network Society, 2nd ed. Oxford, U.K.: Blackwell Publishers.

Cronon, W., ed. 1995. Uncommon Ground: Rethinking the Human Place in Nature. New York: W.W. Norton and Company.

Gallagher, R., and B. Carpenter. 1997. Human-dominated ecosystems: introduction. Science 277: 485.

Giddens, A. 1984. The Constitution of Society. Berkeley, Calif.: University of California Press.

Graedel, T.E., and B.R. Allenby. 2002. Industrial Ecology, 2nd ed. Upper Saddle River, N.J.: Prentice-Hall.

Grubler, A. 1998. Technology and Global Change. Cambridge, U.K.: Cambridge University Press.

Gunderson, L.H., C.S. Holling, and S.S. Light, eds. 1995. Barriers and Bridges to the Renewal of Ecosystems and Institutions. New York: Columbia University Press.

Hacking, I. 1999. The Social Construction of What? Cambridge, Mass.: Harvard University Press.

Harvey, D. 1996. Justice, Nature and the Geography of Difference. Cambridge, Mass.: Blackwell Publishers.

Heidegger, M. 1977. The Question Concerning Technology and Other Essays. Translated by W. Lovitt. New York: Harper Torchbooks.

Hong, S., J. Candelone, C.C. Patterson, and C.F. Boutron. 1996. History of ancient copper smelting pollution during Roman and medieval times recorded in Greenland ice. Science 272: 246–249.

Jablonski, D. 1991. Extinctions: a paleontological perspective. Science 253: 754–757.

Jager, J., and R.G. Barry. 1990. Climate. Pp. 335–352 in The Earth as Transformed by Human Action, edited by B.L. Turner, W.C. Clark, R.W. Kates, J.F. Richards, J.T. Mathews, and W.B. Meyer. Cambridge, U.K.: Cambridge University Press.

Kane, R.H. 1998. The Significance of Free Will. Oxford, U.K.: Oxford University Press.

Landes, D.S. 1998. The Wealth and Poverty of Nations. New York: W.W. Norton and Company.

Lyotard, J. 1979. The Postmodern Condition: A Report on Knowledge. Translated by G. Bennington and B. Massumi. Minneapolis: University of Minnesota Press.

McNeill, J.R. 2000. Something New Under the Sun. New York: W.W. Norton and Company.

Merchant, C. 1995. Reinventing Eden: Western Culture as a Recovery Narrative. Pp. 132–167 in Uncommon Ground: Rethinking the Human Place in Nature, edited by W. Cronon. New York: W.W. Norton and Company.

Michael, D.N. 1995. Barriers and Bridges to Learning in a Turbulent Human Ecology. Pp. 461–488 in Barriers and Bridges to the Renewal of Ecosystems and Institutions, edited by L.H. Gunderson, C.S. Holling, and S.S. Light. New York: Columbia University Press.

Mitchell, W.J. 2000. e-topia: "Urban life, Jim—but not as we know it". Cambridge, Mass.: MIT Press.

Nature. 2003. Welcome to the Anthropocene. Nature 424: 709.

Noble, D.F. 1998. The Religion of Technology. New York: Alfred A. Knopf.

Palumbi, S.R. 2001. Humans as the world's greatest evolutionary force. Science 93: 1786–1790.

Perkins, S. 2003. Three species no moa. Science News 164: 84.

Redman, C.L. 1999. Human Impact on Ancient Environments. Tucson, Ariz.: University of Arizona Press.

Renberg, I., M.W. Persson, and O. Emteryd. 1994. Pre-industrial atmospheric lead contamination in Swedish lake sediments. Nature 363: 323–326.

Rorty, R. 1989. Contingency, Irony, and Solidarity. Cambridge, U.K.: Cambridge University Press.

Rowlands, M. 1999. The Body in Mind. Cambridge, U.K.: Cambridge University Press.

Sagoff, M. 1996. The Economy of the Earth, 2nd ed. Cambridge, U.K.: Cambridge University Press.

Science special report. 1999a. The plant revolution. Science 285: 367–389.

Science special report. 1999b. Genome prospecting. Science 286: 443–491.

Scientific American. 1989. Managing Planet Earth. Scientific American 261(3), (Special Edition).

Scott, J.C. 1998. Seeing Like a State: How Certain Schemes to Improve the Human Condition Have Failed. New Haven, Conn.: Yale University Press.

Senge, P.M. 1990. The Fifth Discipline. New York: Doubleday.

Singer, P. 2001. All Animals Are Equal. Pp. 26–41 in Environmental Philosophy: From Animal Rights to Radical Ecology, 3rd ed., edited by M.E. Zimmerman, J.B. Callicott, G. Sessions, K.J. Warren, and J. Clark. Upper Saddle River, N.J.: Prentice-Hall.

Snow, C.P. 1959. The Two Cultures, reprinted in 1998. Cambridge, U.K.: Cambridge University Press.

Socolow, R., C. Andrews, F. Berkhout, and V. Thomas. 1994. Industrial Ecology and Global Change. Cambridge, U.K.: Cambridge University Press.

Thomas, W.L. Jr., ed. 1956. Man's Role in Changing the Face of the Earth. Chicago: University of Chicago Press.

WCED (World Commission on Environment and Development). 1987. Our Common Future. Oxford, U.K.: Oxford University Press.

The Ethics of Nanotechnology
VISION AND VALUES FOR A NEW GENERATION
OF SCIENCE AND ENGINEERING

GEORGE KHUSHF
Center for Bioethics and Department of Philosophy
University of South Carolina

Big, well funded science needs a vision that can grab the public imagination. For the superconducting supercollider the goal was to discover the fundamental building blocks of the universe. For the Human Genome Project it was to read the book of life. Now the metaphor shifts from discovery to creation, from reading nature to rewriting nature. For nanoscale science and technology the vision involves understanding and manipulating matter at the atomic scale. The vision was described in *Nanotechnolgy: Shaping the World Atom by Atom*, a report by the National Science and Technology Council (NSTC, 1999):

> The emerging fields of nanoscience and nanoengineering are leading to unprecedented understanding and control over the fundamental building blocks of all physical things. This is likely to change the way almost everything—from vaccines to computers to automobile tires to objects not yet imagined—is designed and made.

Obviously, any activity with such huge potential raises a host of ethical and social questions. However, before we can explore these issues, or rather, as a first step in exploring them, we must first clarify what we mean by nanotechnology (Keiper, 2003; Stix, 2001). There are several competing meanings of nanotechnology, and the definition we choose will influence the ethical issues that must be addressed. For this reason, the first part of this essay concerns the debate about how nanoscale science and technology should be understood. I then review the ethical issues that should be considered.

THE MEANING OF "NANOTECHNOLOGY"

"In order to have meaningful discourse on the societal impact of nanotechnology, we must first agree on what we mean by nanotechnology" (Theis, 2001). There are three general approaches to defining nanotechnology. One approach has a very narrow focus but a grand vision; this is Eric Drexler's project of molecular assemblers, or molecular manufacturing. A second approach has an extremely broad focus but no vision; nanotechnology is a grab bag category that includes anything and everything related to the nanoscale, with no significant integrating ideals. The third approach, which has been advanced by the National Nanotechnology Initiative (NNI), attempts to steer a middle way; it focuses on a cluster of factors associated with the nanoscale and articulates a vision of the unique opportunities offered by emerging science and technology. I will argue for the third approach, but first we should appreciate how the other two approaches are shaping and influencing public debate.

Molecular Manufacturing

Eric Drexler coined the term "nanotechnology" in 1986 in his book, *Engines of Creation,* to describe a dramatic new technology—manufacturing at the molecular scale (Baum et al., 2003; Drexler, 1986). Drexler believes that tiny factories could be created that could assemble anything at the atomic scale. These "assemblers" would have molecular computers that would receive blueprints for anything that is naturally possible from which they would then construct these things from raw materials (atoms). Drexler describes assemblers this way:

> Assemblers will be able to make virtually anything from common materials without labor, replacing smoking factories with systems as clean as forests. They will transform technology and the economy at their roots, opening a new world of possibilities. They will indeed be engines of abundance.

In addition to solving environmental problems (for example, assemblers could be dispatched to remove greenhouse gases from the atmosphere and eliminate global warming, inexpensively), these "engines of abundance" could greatly extend human life, solve all energy problems, and enable us to colonize space, to mention just a few of their benefits (Drexler, 2001). There is virtually no limit to what might be accomplished once assemblers have been brought into existence.

Drexler believes their construction would not require any new science but would be simply a large engineering project, akin to landing on the moon, for which all of the basic knowledge is already available (Baum et al., 2003; Drexler, 1992), and he has been trying to convince organizations, such as NNI, to include molecular manufacturing in their research portfolios.

A large group of futurists and technophiles (Transhumanists and Extropians, among others) have adopted Drexler's vision. In their literature, these advocates often link nanotechnology with artificial intelligence. They believe that humans will soon merge with machines, uploading consciousness into computers with vast computational capacity, and that swarms of micro- and nanomachines, such as the "Utility Fog," will lead to smart environments that can change instantly, much like the Holodeck on the science fiction series, *Star Trek*. In their active pursuit of this Brave New World, these advocates use nanotechnology as a buzzword for a radically transformed humanity (Kurzweil, 1999). Another group associated with the Drexlerian vision of nanotechnology, a group radically opposed to it, is composed of people who believe that such technological power threatens humanity with extinction. Perhaps the best representative of this view is Bill Joy, former head of Sun Microsystems. In a widely cited essay in *Wired* magazine, "Why the Future Doesn't Need Us," Joy warns against the potential catastrophe that could result from the convergence of nanotechnology, genetics, and information science (Joy, 2000). One of the risks, he says, is that Drexlerian assemblers will run wild and replicate themselves uncontrollably; using all biomass as raw material, they will ultimately destroy the environment (including all human life). This is the so-called "gray goo problem"—the earth transformed into an indistinct mass of swarming nanobots (Drexler, 1986). To avoid that fate, Joy argues, we must refrain from developing all such technology. Bill McKibben (a noted environmentalist) and Francis Fukuyama (a member of the President's Commission on Bioethics) have joined Joy in calling for a moratorium on using this new technology (Fukuyama, 2002; McKibben, 2003).

The ethical issues raised by nanotechnology understood in its most radical sense—what Drexler now calls "molecular manufacturing"—are framed in grand terms. How can we prevent engines of destruction from reducing the world to gray goo (Freitas, 2001)? How can we ethically navigate a collapse of the world economy that would result from unlimited production by assemblers (Phoenix and Treder, 2003)?

Several groups have been formed to address these issues. The Foresight Institute, founded by Eric Drexler and Christine Peterson, regularly sponsors workshops to address the ethical and social impact of nanotechnology. This group has formulated guidelines for the development of nanotechnology that would minimize its adverse impacts (Foresight Institute, 2000). The Center for Responsible Nanotechnology (2002), headed by Mike Treder and Chris Phoenix, focuses on anticipating radical transformations and providing guidance for the new economic and legal order that will follow. In addition, some lawyers, such as Glen Reynolds, are considering the legal issues that might be associated with nanotechnology (Forest, 1989; Reynolds, 2001, 2002). All of these groups and individuals are sympathetic to the general goal of molecular manufacturing. In fact, they celebrate that goal and wish to see it actively advanced. To allay the fears of people like Bill Joy, however, they attempt to show how nanotechnology can be developed responsibly.

Other groups, like the ETC Group (an environmental organization that has been influential in keeping genetically modified organisms out of Europe), share a similar view of nanotechnology but are opposed to its development. They hope that research will be halted altogether so that a more natural world can be preserved (ETC Group, 2003). By contrast, the environmental group Greenpeace is much more skeptical about the whole idea of assemblers (Arnall, 2003).

Whenever the focus of the discussion is on the radical implications of nanotechnology, the debate on ethical and social issues takes on a grand tone, similar to the tone of the debate about nuclear reactors or genetic engineering. Issues are framed in visionary terms, with an unavoidable sense that we are dealing with a new world order. Framing the debate in this way has some advantages, because no matter how one understands nanotechnology, everyone appreciates that it is likely to have radical, long-term effects, and it is important that we try to anticipate them and respond accordingly.

There are also significant disadvantages to framing the debate this way. The Drexlerian vision, although it is influential, does not address the great majority of research being done under the heading of nanotechnology. Only one company— Zyvex—has a stated goal of creating assemblers (Ashley, 2001), and many view this project with considerable skepticism. In fact, many scientists consider Drexler-type molecular manufacturing science fiction.

In one very visible debate, Richard Smalley, Nobel Laureate in chemistry for the codiscovery of C-60 (fullerenes), argues that Drexler is "in a pretend world where atoms go where you want because your computer program directs them to go there." He accuses Drexler of not appreciating basic concepts and constraints associated with chemistry (Baum et al., 2003; Smalley, 2001; Whitesides, 2001). Some may think Smalley is a bit unfair, and on one level the debate could be seen as a squabble between disciplines that converge in the nanotech arena, with Smalley on the side of chemistry and Drexler and associates, like Merkle (formerly with Zyvex), on the side of computer science and systems engineering.

But even if Drexler's vision is given a more sympathetic reading, his proposals must be considered speculative, and what he means by molecular manufacturing must evolve considerably before anything like it can be approximated in practice. (One already sees such an evolution of the concept in the way Zyvex conceptualizes its core goals.) The whole project of assemblers is still very much outside the mainstream of current research, and it would be unfortunate if the primary debate on nanotechnology were closely associated with developments that are, at best, on the periphery of what is actually being done by scientists.

So the molecular-assemblers definition can be summed up as follows. Drexler's vision is influential and has a high public profile. When the public hears about nanotechnology, it will probably be through movies like *Agent Cody Banks*, in which a secret agent has to protect the world from a deranged megalomaniac who wants to unleash self-replicating nanobots, or Michael Crichton's novel, *Prey* (2002), in which a Zyvex-like company called Xymos, originally funded by the Defense Advanced Research Projects Administration (DARPA), unwittingly releases nanoswarms that evolve toward a similar, destructive end. Thus, when the public thinks about nanotechnology, it is likely to be in Drexlerian terms.

In addition to popular entertainments, people who have testified before Congress and who are often cited in media reports on nanotechnology are also associated with molecular manufacturing. Of course, we must be aware of this debate, and we must understand how the ethical issues are therein addressed. But, in the end, this is a small, marginal area of research in nanotechnology, and the ethical issues are much broader than this debate would indicate. In fact, molecular manufacturing—understood as a kind of directed, positional assembly of anything—is much too narrow a definition of nanotechnology, and the ethical issues associated with this definition are, at best, a subset of the broader issues.

A Grab Bag of Unrelated Research

In recent issues of *Smalltimes*, a journal associated with nanotechnology, there was an interesting debate on the meaning of the term. Ken Galleo (2003), a technologist at Cookson Electronics, wrote an open letter asking for clarification (see also Mickelson, 2003). Galleo notes that "nano definitions are murky and numerous, and those [like Drexler's] that exclude mass chemical and bio reactions as too imprecise and random seem exceedingly limiting." Smalley and other prominent researchers engaged in nanotechnology projects advance a notion of nanotechnology that goes far beyond the idea of assemblers. However, if one takes into consideration mass reactions (like those that take place in chemistry generally), it is difficult, if not impossible, to distinguish nanotechnology from other things that are in some way related to events at the nanoscale (and isn't that everything?). Galleo concludes:

> Web sites, especially governmental [web sites], slip past a real definition to quickly praise nanotechnology without explaining terms and what they want to sponsor. No wonder we're seeing articles about "nano-pretenders" and "nano-hoaxes." So before the nanotechnology definition lapses into "anything less than 100 nm," can we please get a better definition?

Anthony Vigliotti (2003) responded in a later issue of *Smalltimes* with a letter given the headline of "A No-Nonsense Nano Definition." Vigliotti compares definitions of nanotechnology to glitter at a birthday party, "the definitions are sparkling and exciting, yet annoying and quickly thrown in the trash." He goes on to suggest that "the definition should be as small as possible and written like it came out of the dictionary of 2053, when the technology should be commonplace." His proposed definition is "the creation and exploitation of 1 to 100 nm structures."

Many others who have advanced a minimalist definition of nanotechnology have skeptical or cynical reasons for doing so. Even some directors of nanocenters and prominent researchers in the field have opined, often with a whisper and a wink, that there is nothing more to nanotechnology than a general focus on scale (roughly 1 to 100 billionths of a meter). According to this definition, nano-technology includes a host of diverse technologies and research endeavors, such as catalysis, molecular electronics, and new nanopharmaceuticals (just to mention a few), that are, at best, distantly related. The members of this heterogeneous group have little in common other than the scale of some components.

According to skeptics, these research topics are grouped together solely for funding purposes. "Nanotechnology is really a convenient label for a variety of scientific disciplines which serves as a way of getting money from government budgets" (Doug Parr, in the Foreword to Arnall, 2003; Roy, 2002). Government and industry only come up with substantial funding for research when the sub-ject is new and "hot," and social and cultural forces have made nanotechnology a convenient label for lobbying and funding efforts. Benefiting from the hype associated with the science-fiction-like powers of assemblers, nanotechnology has become a catchphrase for "great new science with lots of promise." In fact, a more precise definition than "1 to 100 nm" would actually exclude some research areas and could start a turf war that most researchers would rather avoid. Under careful scrutiny, people might discover that the emperor is not wearing any clothes.

This skepticism and cynicism may have some basis. As Galleo notes, the more you read about nanotechnology, "the less clear and more ambiguous the meaning becomes." This is partly because nanotechnology is the new area for megafunding. In 2003, $774 million in federal funds was allocated ($64 million more than the projected $710 million); for the fiscal year beginning in October 2003, the projected amount is $849 million (Roco, 2003a). Some researchers have creatively redefined their projects so they can qualify for funding associated with nanotechnology. In fact, it is hard to see how these diverse research endeav-ors can be included under a single heading. Nanotechnology has become a grab bag for loosely related science and engineering projects that focus on the nano-scale. Although most of these projects are valuable, they are very conventional.

For scientific reasons, the minimalist definition will not do. Whereas Drexler's approach is futuristic, narrow, and disconnected from current science, the minimalist definition is mundane and much too broad. If nanotechnology simply concerns the creation and exploitation of 1 to 100 nm structures, then, as Paul Alivisatos (2001) notes, "All of biology is arguably a form of nanotech-nology." In addition, most of chemistry instantly becomes nanotechnology, as does a great deal of materials science, physics, and so on.

Although it is difficult to pinpoint, something unique and exciting is emerg-ing in nanotechnology, but putting that something into words is more of a philosophy-of-science project than a science project per se. It is a metascientific endeavor important for scientists because it will facilitate the development of that emergent something.

A clear definition is also important for addressing ethical and social issues. If nanotechnology is nothing more than a grab bag for a host of unrelated projects, then the ethics of nanotechnology becomes nothing more than the ethics of unrelated projects taken individually, or the ethics of science and engineering in general. Of course, a host of ethical issues are associated with science and engineering, including research integrity, workforce and product safety, and the impact of new products on society, just to mention a few. The question of "hype and funding" (Arnall, 2003; Roy, 2002), how socioeconomic factors affect the configuration of research enterprises, will also be important. Of course, all of these issues will be part of a nanoethic, no matter how nanotechnology is defined.

The basic question remains, however. Are there unique ethical issues associated with nanotechnology? If so, what are they? We cannot really answer this question until we determine if nanotechnology is in some way unique, and, if so, how its uniqueness can be characterized. In that sense, an appropriate characterization/definition of nanotechnology is an important preliminary to a discussion of nanoethics.

The Middle Way

In general, it is difficult to define a phenomenon until it has reached maturity. In the words of G.W.F. Hegel, philosophy (and science—Hegel equated their logic) can only paint its conceptual gray on gray after a form of life has grown old. Only then, at dusk, does the Owl of Minerva take flight. Dusk is the time of definitions. However, "nanotechnology is still in its infancy" (Roco and Bainbridge, 2001). At dawn, when ideas first struggle forth, there is always a tangle of science and fiction, vision and value, thought and feeling. The richest ideas often emerge as apparent contradictions, strange juxtapositions of future and present.

Perhaps this is a time for characterization, rather than definition. Characterization can provide content and coherence and can define the scope and range of issues, but not an identification of necessary and sufficient conditions. Beyond a description of what already exists, a characterization can direct our gaze toward the future and suggest a shape that can only be seen faintly and with great effort. By providing coherence that is not yet fully there, characterization itself becomes a moment in the process of formation, in this case, a moment in the development of nanotechnology.

Thus, characterization is both descriptive and constructive, capturing where nanotechnology is now and where it should go. The word "should" has both an ethical and a scientific component. Where will this science take us? What will be its form, and how will the body of knowledge be structured so the world is appropriately known and we are situated to intervene? These questions cannot be answered without scientists and engineers. But answering them requires an act of will, a decision about where we should go and what we should be.

This new knowledge, with its tremendous capacity to alter our landscape permanently, is intimately intertwined with values, which cannot be fully disentangled from questions of science. How should we shape our future? What social institutions should be put into place, and how should the public participate in the formation and use of this technology? These are questions of ethics and social policy. The initial characterization of nanotechnology must include all of these considerations, which heralds a radical change in the way we address ethical issues.

Traditionally, we have assumed a kind of linear development from science to engineering—first knowledge, then the application of such knowledge to advance human ends. Ethics and values only came in at the second step, in assessing the uses and abuses of scientific knowledge. That model is no longer satisfactory.

In the realm of technoscience, fact and value are intertwined, as are the basic and applied domains of science. As Roco and Bainbridge (2001, 2002) note, nanoscale science and technology are "at the unexplored frontier of science and engineering," and both science and engineering will be fundamentally transformed as a result. The broader relationship between science, engineering, and ethics will also be transformed.

Science and ethics can no longer relate in a two-step process. Each informs the other, playing a co-constructive role in the process by which a new science and technology, such as nanotechnology, evolves (Weingart, 2002). The ethics of nanotechnology belongs in this richer, collaborative context of sciences and humanities. The first step in defining that ethic is to characterize what exactly is at issue. Characterization in this context is formative and constructive, not an act that can be done once and for all. It is an ongoing process that must attend the development of the science.

The first step in that characterization has been taken by the leaders of NNI. Nanotechnology is defined in *National Nanotechnology Inititiative: The Initiative and Its Implementation Plan*, issued in 2002 (NSTC, 2002):

> The essence of nanotechnology is the ability to work at the molecular level, atom by atom, to create large structures with fundamentally new, molecular organization. Compared to the behavior of isolated molecules of about 1 nm (10^{-9} m) or of bulk materials, the behavior of structural features in the range of about 10^{-9} to 10^{-7} m (1 to 100 nm) exhibit[s] important changes. Nanotechnology is concerned with materials and systems whose structures and components exhibit novel and significantly improved physical, chemical, and biological properties—and that enable the exploitation of novel phenomena and processes—due to their nanoscale size. The goal is first to exploit these properties by gaining control of structures and devices at atomic, molecular, and supramolecular levels and then to learn to manufacture and use these devices efficiently. Maintaining the stability of interfaces and the integration of these "nanostructures" at micron-length and macroscopic scales are all keys to success.

New behavior at the nanoscale is not necessarily predictable from that observed at larger size scales. . . . Being able to reduce the dimensions of structures down to the nanoscale leads to the unique properties of carbon nanotubes, quantum wires and dots, thin films, DNA-based structures, and laser emitters. Such new forms of materials and devices herald a revolutionary age for science and technology, provided we can discover and fully utilize the underlying principles.

Although this is called a "definition," it would be more accurate to call it a characterization, because it does not identify the necessary and sufficient conditions for an object or activity to be counted as nanotechnology.

It does provide a useful description of nanotechnology. It captures the idea that at the 1 to 100 nm scale, novel properties emerge. The task of nanoresearch, then, is to discover these properties, learn to control their expression, develop the tools for scaling them up to microscales and macroscales, and then develop manufacturing on a large scale (Roco et al.,1999). If successful, the results would lead to "a revolutionary age for science and technology."

The characterization also raises several questions. What accounts for the unique properties? Why do they emerge, and what are they? How is science altered, and what is revolutionary about this? Are these claims inflated, or is there something qualitatively different about what happens on the nanoscale? A detailed discussion of these questions and of the features identified in the NNI definition is far beyond the scope of this paper, but we can briefly touch on some of them to get a sense of the stakes in nanotechnology (Khushf, 2004b).

The Mesorealm

The nanoscale bridges quantum and classical effects. At the bottom end of the scale, quantum effects dominate; at the top end, classical effects dominate. Many of the interesting properties associated with nanotechnology exist in this strange middle world, the mesorealm, where, as Michael Roukes (2001) notes, "unforseen properties of collective systems emerge." These properties include the relationship between the size of a quantum dot and the wavelength of light it emits and the quantum character of thermal or electrical conductivity on the nanoscale.

A Bridge between Physics, Chemistry, and Biology

Developments in the natural sciences have converged at the nanoscale. "At the nanoscale, physics, chemistry, biology, materials science, and engineering converge toward the same principles and tools" (Roco and Bainbridge, 2001). Thus, the metaphors used to describe the relations between these sciences have changed. In the past, hierarchical metaphors were used. Physics was understood to be the base; chemistry was built upon that base; and biology drew upon both physics and chemistry. The grand goal, the unity of science, involved reducing the higher levels of the hierarchy to the lower level. Ultimately, everything was to be understood in terms of, and translated into, the terms of physics, the most foundational science.

With nanotechnology, the relationships between the sciences are more symmetrical. Biology is still informed by physics and chemistry, but biology and medicine have taken a "molecular turn," with revolutionary implications for the future of both. Physicists and chemists also look to biology, not just for applications, but also for a better understanding of fundamental science in their own domains. The neat distinctions—between organic and inorganic chemistry, between living and nonliving systems, and between the natural and the artificial worlds—have been blurred (Buchand and Montemagno, 2000; Goldstein, 2003a, 2003b; NSTC, 2002; Roco, 2003b). The metaphors of hierarchy and reduction have changed to a metaphor of bridging.

Tools for Visualization and Manipulation in the Nanorealm

In essays often regarded as founding documents for the field of nanotechnology, Richard Feynman (1992, 1993) said one of the most important things that can be done to advance biology, and the broader project of scaling down in all areas, was to improve the resolution of the electron microscope. This has, of course, been accomplished; the electron microscope is now capable of "seeing" even beyond the low end of the nanoscale.

For important reasons, the scanning probe microscope, rather than the electron microscope, is now a significant icon of nanotechnology (Baird and Shew, 2002). The use of such a microscope for the directed manipulation of atoms (to write "IBM" with xenon atoms, for example) is a paradigm of the potential of nanotechnology (NSTC, 1999). In this example, "seeing" and "acting" merge in complex ways, much as they do at the quantum level, with as yet unexplored implications for the meaning of both "visualization" and "manipulation." Multiple strategies, from various forms of microscopy to x-ray crystallography to theoretical and computational tools, are now used to understand structure and function. Overlapping imaging techniques are often used to produce a single image; thus, seeing merges with constructing on multiple levels, knowledge merges with doing (Baird, 2004).

Strategies for Manufacturing

At the top end of the nanorealm, some manufacturing strategies simply scale down macro- and microstrategies; for example, in photolithography, an image of a desired pattern (e.g., a silicon chip) is projected and etched. At the bottom end of the nanorealm, some chemical methods work through mass action, and with quantities assessed in moles, to form new kinds of molecules. The challenge for nanotechnology has been to develop methods that link bottom-up, mass-action strategies with top-down, directed manufacturing.

At this point, much of the research is still seen in terms of alternatives—either more practical, top-down strategies or bottom-up, self-assembly strategies. But the key challenge involves a merging of chemical self-assembly and systems engineering strategies that will create new interdisciplinary boundaries and new conceptions of both the theory and practice of manufacturing (Whitesides, 2002; Whitesides and Love, 2001).

CENTRAL FEATURES OF NANOTECHNOLOGY

These scientific considerations must be part of our ethical deliberations because they are linked to the kinds of properties that will emerge, the principles that will be used, and the tools for manipulation at the molecular scale. Until we know the properties and how they arise, we cannot assess risks or contemplate how we should intervene. Broad concerns must be confronted—how living and nonliving systems will be linked and how we will "see" and "act" in the nano-realm. To ignore these concerns would be like ignoring the details of embryology in the stem-cell debate or ignoring the mechanisms of heritability in genetic engineering. The concepts of manipulation and self-assembly (just to give two examples) are not purely scientific. They can be scaled up to the macrolevel (but often not directly because they have only analogous links to common-sense notions of these concepts); thus, they interface with broader ethical concepts. The debate about the meaning of these terms will thus have a social/values component that should be made explicit even at this early stage in the discussion.

In addition to the core scientific considerations, there are also certain characteristics of interfaces between diverse disciplinary sciences, science and broader engineering projects, and science and social policy. The following characteristics are often considered central to nanotechnology.

Nanotechnology Is Fundamentally Interdisciplinary

Entirely new facilities are being designed to support and foster the interdisciplinary possibilities of nanotechnology. In addition, there are obvious implications for the education of new scientists and engineers and the allocation of resources to establish the necessary infrastructure to sustain nanoresearch. These workforce and infrastructure issues have an obvious societal dimension; in fact, this is one of the social implications addressed by NNI (NRC, 2002; NSTC, 2002; Roco and Bainbridge, 2001, 2002).

The interdisciplinary nature of nanotechnology also has serious implications for the social configuration of scientific and engineering practices. Issues related to interdisciplinary pursuits and projects are notoriously difficult to address, especially in university settings, where the mechanisms for research and knowledge dissemination are closely intertwined with disciplinary identity (NRC, 2002). Sustained efforts and careful reflection will be necessary to "establish a shared culture that spans across existing fields" (Gorman, 2002).

Nanotechnology Challenges Distinctions between Pure and Applied Science

In the literature about nanotechnology, the common distinction between fundamental science and engineering is still taken for granted, but it has been relativized. Clearly, the older boundaries between the pure and applied domains can no longer be sustained (Roco and Bainbridge, 2001, 2002). In fact, this is now taken for granted in the way big science is funded (Etzkowitz, 2001). We no longer assume that the "pure sciences" will pursue "pure knowledge," with application following naturally as a second step. Instead, various goals are advanced, and the fundamental research likely to advance those goals is supported. For a successful grant application, documentation of the practical applications of research results is essential, even for basic science research.

However, the blurring of boundaries goes far beyond funding criteria. In the past, it was assumed that science was purely fact based, and thus independent of broader socioeconomic concerns. Today, it is broadly recognized, especially in the philosophy-of-science community, that the independence of science is, at best, a regulatory ideal. The human ends that guide engineering applications unavoidably reflect back upon fundamental scientific research, even in the purest areas of science. This means that the values intertwined with those ends also play a role in the science (Weingart, 2002).

In nanoscience, the link between fact and value is increasingly explicit. Thus, ethics, with its disciplined reflection on values, goes to the very root of this technoscience. Ethical considerations must be addressed, even in the earliest stages of research. And they must be addressed explicitly, rather than implicitly.

Nanotechnology Requires a Framework for Integrating across Scales

Mike Roco, the director of NNI and one of its architects, and William Bainbridge, the deputy division director of the Division of Information and Intelligent Systems, National Science Foundation, have said that achieving the goals of nanotechnology will require "nothing less than a fundamental transformation of science and engineering" and that one of the "substantial intellectual barriers" for accomplishing this transformation involves the "development of a hierarchical architecture for integrating sciences across many scales, disciplines, and data modalities" (Roco and Bainbridge, 2002).

By "hierarchical architecture" they do not mean the traditional hierarchy of disciplines associated with the older, reductionist view. They are looking to systems theory or some other holistic framework to provide guidance for linking across scales. This framework must do no less than bridge the cultural divide between science and the humanities. It must link the activities of scientists and engineers with the human goals and social framework that will be radically influenced by nanoscience and engineering projects.

Ethical/Social Considerations Are Basic Features of Nanoresearch

For the reasons outlined above, and for many reasons not directly addressed here, nanoscience cannot be based on traditional models in which ethical/social reflection is a second, later step in the assessment of the use and/or abuse of previously configured science. Ethical reflections must accompany research every step of the way, and this should be a defining feature of nanotechnology, not just a statement about how ethical issues should be addressed. Indeed, there is a realistic possibility that this will happen. "As the NNI is commencing, there is a rare opportunity to integrate the societal studies and dialogues from the very beginning and to include societal studies as a core part of the NNI investment strategy" (Roco and Bainbridge, 2001). Clearly, the architects of NNI envision the emergence of new kinds of science and engineering and with them a new way of interfacing with society.

LEVELS OF ETHICAL REFLECTION

Now that we have some sense of what nanotechnology is about, we can consider the ethical issues associated with it more explicitly. Three levels of ethical analysis are required: (1) critical reflection on the vision and values of nanotechnology; (2) the formation of ethical theory; and (3) specific topical areas.

Critical Reflection on the Vision and Values of Nanotechnology

The first level of ethical analysis was included in the discussion of characterizing nanotechnology. Through critical reflection on the scientific and engineering enterprise itself, the values and core ideals of nanotechnology can be specified so that sustained ethical reflection can accompany research even at the preliminary stage. This is probably the single most important development to ensure the ethical integrity of nanotechnology. If an ethical awareness and culture become part of the research enterprise, then all of the other components of ethical analysis are likely to fall into place. If not, nanotechnology is likely to struggle with the same polarization of scientific and ethical analysis that has plagued other controversial areas, such as nuclear technology and genetically modified organisms.

A culture of ethical awareness has several prerequisites. One of the most important involves bridging the cultures of scientific practice and ethical analysis. The difficulty of accomplishing this should not be underestimated; it will require bridging the so-called "two cultures divide" that has characterized not just the academic arena, but also society as a whole. Ethical analyses are inextricably intertwined with the historical, philosophical, social, and religious narratives of diverse communities. Differences among these communities are navigated in complex ways, and when the values of other communities are perceived to impinge upon scientific research and engineering practices, scientists often regard them as unwanted and inappropriate intrusions.

Broad historical narratives, such as those associated with Galileo and Darwin, are paradigms of how unwanted influence can distort the character of a science. Scientists often attempt to address ethical issues associated with their research in secrecy and independently of the "uneducated" public eye, and they tend to resent those who bring close public scrutiny to these issues. To move beyond these antagonisms, we will need new models for constructive interaction between the humanities and the sciences that avoid distortions of science, as illustrated in the Galileo and Darwin case studies, but that encourage appropriate ethical and social input in areas of science that have a clear values component and huge social implications (Weingart, 2002).

Scientists must appreciate how broader kinds of human discourse guide ethical analysis; and people in the humanities must learn as much as they can about the relevant science. "Trading zones" could be established to provide a venue where these two communities can learn each other's languages and acknowledge each other's interests (Galison, 1997; Gorman, 2002). This will not happen in a few weeks or months, even with concerted effort. As it has in similar situations, such as medicine and bioethics or the environment and environmental ethics, it will take time to develop nanoethics. Researchers at this intersection must make ethical issues and problems a core part of their formal activities (Weil, 2001).

People on all sides must recognize that it will take time and patience for a dialogue to emerge. Those who specialize in ethics and the humanities, on one side, and those who specialize in science and engineering, on the other, will have to learn to respect each other.

Nanoresearchers must consider ethical issues part of their research per se. As specialists emerge, ethicists should be integrated into their research teams, especially in controversial areas of investigation. At this early stage of research in nanotechnology, we should be asking how we can foster an ethical culture and how we can train both ethicists and scientists. We need an infrastructure and educational initiatives not only to establish the next generation of nanoresearchers, but also to establish appropriate ethical analysis. Now is the time to ask how we can accomplish these goals.

The Formation of Ethical Theory

There will be many areas where ethical approaches developed in other contexts can be directly applied to nanotechnology, but it will also be necessary to develop forms of ethical reflection that are more closely wedded to nanotechnology. These forms of "situated" ethical reflection will emerge as the field matures and researchers are compelled to deliberate on the ethical issues that arise.

A helpful example can be found in the history of biomedicine. When bioethics first emerged in the 1950s and 1960s, the divide between the humanities and biomedicine was similar to the current divide between the humanities and the nanosciences. On one side, there was a professional culture of medicine and biomedical science that had its own forms of ethical reflection, often as much etiquette as ethics related to paternalistic attitudes toward patients and the larger public. The ethical norms of the medical/biomedical community were largely implicit and did not address a host of concerns considered important by people outside the profession. On the other side, there were rich traditions of theological and philosophical ethical reflection, but those trained in these areas had little understanding of medicine or the biomedical sciences. Therefore, when they attempted to apply their theories, they were often extremely insensitive to the realities of clinical and scientific practice.

As bioethics matured, forms of middle-level analysis were developed that bridged the realities of the clinical and scientific world and the world of ethical reflection (Khushf, 2004a). Ethical theories, such as principalism and casuistry, were formulated in ways that were specifically oriented toward biomedical applications, and as knowledge of these theories was disseminated, a shared culture of ethical reflection emerged that directly incorporated clinical and ethical expertise.

Although the ethics of biomedicine still has deficiencies, and it would be difficult to transfer the process that led to bioethics to nanotechnology, I do think there is a valuable lesson here and that we should look for a similar kind of development. We can expect that as those with broad experience in ethics and social/cultural analysis enter into a dialogue with researchers in nanoscale science and technology, strategies of ethical analysis will emerge that will be oriented toward the specific needs of nanotechnology.

One area where nano-specific ethical theory will be needed—risk analysis—is already apparent, and bioethics offers a useful analogy. Risk analysis involves a complex integration of risk assessment, risk management, and communication. Risk analysis is widely used to manage uncertainty surrounding the introduction of a hazardous chemical or a large-scale engineering project; it is also integral to the way we determine whether a given research protocol can be advanced in human subjects or whether a given treatment option ought to be pursued for a patient (National Commission for the Protection of Human Subjects, 1978). Risk analysis is often intertwined with more general economic assessments; in those cases, utilitarian theory (which involves some accounting of the goods to be realized and the harms to be avoided) is used to frame the analysis.

All such forms of risk analysis interface in complex ways with the particular visions of those who might be influenced by the proposed intervention, and various theories (such as stakeholder theory in business ethics and principalism in bioethics) have been advanced to consider how stakeholders might participate (Beauchamp and Childress, 2001). At the same time, theories of the market and free exchange, as well as more directive theories of governmental oversight, also address how risks should be assessed (Engelhardt, 1996). In the biomedical arena, many of the most widely used forms of risk analysis address a balance between autonomy (when the individual influenced most by the decision gets to decide) and beneficence (when decisions are based on broader assessments of overall utility). This balance often depends on prudential deliberation that cannot be captured in systematic rules. Specific medical ethics doctrines, such as informed consent, preserve the balance between autonomy and beneficence (Beauchamp and Childress, 2001).

Nanoethical strategies analogous to bioethical and enviroethical strategies will have to be developed to guide risk assessment (Stuart, 2003a,b). Some early attempts at addressing the risks of nanoscience by Greenpeace (Arnall, 2003) and ETC Group (2003) simply apply the so-called "precautionary principle," which states that those who would introduce a new product must show that it is safe. As a framework for the early debates on genetically modified organisms, this approach resulted in their exclusion from most European markets, and the debates were divisive and polarized. For reasons that cannot be addressed here, I believe that the precautionary principle would be inappropriate as a basis for debates on nanotechnology. (I also think it is inappropriate for debating environmental technologies, but that is a separate issue.) In fact, calls for a moratorium based on the precautionary principle make no sense, because they do not take into account the diverse and complex kinds of research that would be affected or the stages of development in the research (Wardak and Rejeski, 2003).

However, traditional utilitarian analyses will not be appropriate either. It will take a good deal of groundwork to identify the kinds of risk involved, determine how these might be assessed, and determine how objective and subjective assessments are incorporated into broader decisions about implementing new programs, especially controversial programs that affect human performance (Smith, 2001; Weil, 2001).

The problems associated with risk analysis and guiding principles, such as the precautionary principle (or some alternative), highlight just one area in which a nanospecific ethic will be needed. At this stage, efforts should be concentrated on identifying areas that will require sustained reflection. It should also be recognized that much of the ethical theory will only emerge later. Nanoethics, like bioethics, is likely to emerge only after core topical areas have been identified and addressed.

Specific Topical Areas in Nanotechnology

The embryonic stage of nanoethics is apparent when one compares the explosive increase in research with the minimal increase in attention to ethical reflection (Mnyusiwalla et al., 2003). The few studies that have been done on ethical issues have focused on identifying topics that should be addressed (Roco and Bainbridge, 2001). Some have attempted to apply general engineering ethical concerns, such as the integrity of research or fairness in distributing benefits, to nanoscience. As Mnyusiwalla et al. (2003) note, ethical deliberations have not moved beyond "generalization and motherhood statements." Indeed, we have not reached the stage at which we can move beyond a general geography of the ethical landscape, a necessary first step, partly because we have yet to identify and characterize the core areas of nanoscience satisfactorily. As topics are identified, we can begin addressing them in greater depth.

One strategy for identifying core areas involves distinguishing between ethical issues related to particular subtopics of nanotechnology and ethical issues related to nanoscale science and technology generally. Ethical issues related to a subarea of nanotechnology will be closely intertwined with the details of specific research projects and can be addressed as a component of the research; perhaps an ethical/social issues module can be incorporated into grant applications or private funding initiatives (Weil, 2001). Some examples of a few subareas in nanotechnology that might be considered are given below.

Catalysis

Catalysis, which has been called "old nanotechnology," is a traditional focus of chemistry and plays a large role in the economy. New imaging tools, coupled with advancements in bottom-up methods (associated with self-assembly) and careful control of synthesis, has led to unprecedented control of chemical processes, with significant implications for "better, cleaner, and more capable industrial processes" (NSTC, 1999). Ethical reflection in this area should be focused on identifying new developments, especially in revolutionary areas, such as energy and the environment, and their social implications. Weil (2001) identifies catalysis as one of the concrete areas that "can provide points of entry to the institutional, organizational settings in which potential [ethical] problems are embedded and in which they must be examined."

Molecular Computing

Molecular computing, one of the most active areas of research in nanotechnology, also has some revolutionary implications (Lieber, 2001). Molecular computational capacity could lead to computers the size of sugar cubes that are millions of times faster than today's computers. These tiny computers could potentially change every area of life by changing the interfaces between humans and their environment. "Everyone and everything conceivably could be linked all the time and everywhere to a future World Wide Web that feels more like an all-encompassing information environment than just a computer network" (NSTC, 2002). The implications involve privacy, the economy, communications, public policy, even how we understand ourselves as social beings.

Nanomaterials

Materials with new properties can be built on the basis of nanotechnology. Nanocomposites will enable the construction of lighter cars and planes, which would greatly reduce energy consumption. Nanofiltration systems could address environmental problems. In addition to technologies that could improve sustainability, there may also be new risks, such as toxic health effects or adverse environmental effects. The focus should be on identifying new materials, new properties, and new products so that their ethical and social implications can be addressed early on.

Military Applications

The Army has provided the Massachusetts Institute of Technology with $50 million to develop nano-based technology to equip future soldiers. The key areas of research include: "threat detection, threat neutralization, concealment, enhanced human performance, real-time automated medical treatment, and re-ducing the weight load of the fully equipped soldier" (NRC, 2002). In addition, the DARPA funds a great deal of research related to nanotechnology, which is likely to have a huge impact on the military (e.g., on human-machine interfaces). Military uses of nanotechnology should be the subject of careful ethical analysis, not only because they will affect military personnel, but also because they are likely to be transferred quickly to nonmilitary settings. One need only consider DARPA's funding of the Internet to imagine the potential impact of such technology. We also need to consider how nanotechnology might be used by hostile groups, including terrorists, and how it might be used to improve global security.

Space Applications

As Samuel Venneri (2001), chief technologist at the National Aeronautics and Space Administration (NASA), has noted, "nanotechnology encompasses the attributes of self-generation, reproduction, self-assembly, self-repair and natural adaptation. These are all attributes we attribute to living things. . . . Nanotech-nology will enable NASA to build future systems with many of these 'life-like' characteristics." Such developments are needed so we can travel great distances from Earth and withstand the harsh environments that will be encountered in space. Venneri also notes that building things with these lifelike characteristics will challenge NASA's traditional ethics. Many of the issues, he says, "are moving beyond the typical bounds of technology into the domain of natural philosophy."

Biomedical Applications

Nanotechnology has the potential to transform medicine, enabling new diagnostic and therapeutic capabilities that could "fundamentally alter patient-doctor relationships, the management of illnesses, and medical culture in general" (Alivisatos, 2001; NSTC, 1999; Roco, 2003b; Smalltimes, 2003). In addition, nanotechnology could greatly enhance human performance by slowing aging processes, providing new sensory capabilities, and enabling direct brain-machine interfaces (Fukuyama, 2002; McKibben, 2003; Mnyusiwalla et al., 2003; Moore, 2003; Roco and Bainbridge, 2002). Alan Goldstein, the director of biomedical materials engineering at Alfred University, has expressed serious concerns about these developments: "Even at this primitive stage, bioengineering creates a startling constellation of ethical considerations; for the patient's humanity, for healthcare policy, and society. The need to integrate technology and ethics will only increase in scope and significance as the field becomes more mature. Enabled by nanotechnology, bioengineers will soon be integrating neurons with diodes, DNA with transistors" (Goldstein, 2003a,b). Ethical issues at the nano/bio interface will require intensive research and reflection.

Energy

Nanotechnology has great potential to address energy and environmental problems. Examples include high-efficiency fuel cells, artificial photosynthesis, new catalysts, and technologies for reducing energy consumption. Despite its great potential, some environmental groups have already compared nanotechnology to nuclear energy in terms of promise and risks (Arnall, 2003; ETC Group, 2003). This is why it is important that we address the ethical issues early in a broad public debate, not just in terms of promise and risk. We will need a constructive dialogue about the promise of nanotechnology for addressing energy needs before attitudes based on old ideas and insufficient information have been developed and become embedded. Environmental groups are not the only groups that may be skeptical about nanotechnology. The energy industry (particularly the oil industry) may be opposed to the development of nanotechnologies for economic reasons.

Other Areas for Ethical Consideration

Many general ethical/social concerns not related exclusively to nanoscale science and technology pervade all of the topical areas mentioned above. These include: workforce and education; environmental and health impacts; the nano-divide; commercialization and funding; privacy and security; general cultural and societal impacts; the form of public debate; images in science, science fiction, and the media; and legal issues. Extensive work will be necessary in all of these areas.

We already have a sufficient knowledge base for developing summary documents in many of these areas, so we can begin to situate nanotechnology in the context of the larger ethical and social debate. But first we must determine the degree to which nanoscience is unique in that it raises specific problems that require specific solutions and the degree to which it can be considered an instance of the ethical concerns associated with all sciences and technologies. This will require a dialogue between those working in nanoethics and those working on the general ethics of science and engineering. Some of the work on engineering ethics, for example, ethical considerations associated with a culture that sustains research integrity, can surely be integrated into nanoethics. To this extent, at least, researchers in engineering ethics, business ethics, environmental ethics, and bioethics must participate in this dialogue.

Another strategy for organizing and addressing topical areas in nanotechnology is a timeline. A recent conference on the societal implications of nanotechnology was based on this strategy. The conference "propose[d] a vision and alternative pathways toward that vision integrating short-term (3 to 5 year), medium-term (5 to 20 year), and long-term (more than 20 year) perspectives" (Roco and Bainbridge, 2001). Richard Smith (2001) showed how a timeline could be used to address ethical issues; for each time period he characterized what nano-technology would entail and the kinds of problems that were likely to be raised.

In Smith's account, in the short term, nanotechnology will be mostly in the research phase; some microelectromechanical systems (MEMS) and nanosensors will be tested and deployed, and various coatings and materials will be nearing the final stages of development. Ethical and social issues will revolve around the education of the next generation of researchers; commercialization; preliminary risk assessment; the use of terms, such as "nanosystems"; interdisciplinary problems; funding priorities (especially for visionary kinds of nanotechnology); and international competition and cooperation. In the short term, funding, research, and focus will be "widely dispersed politically, geographically, technically, and scientifically."

In the midterm, Smith believes nanotechnology will involve super-MEMS and "entirely new classes of materials and manufacturing processes" that will become part of our everyday lives. New, nano-based diagnostic systems will be available, communicating and programmable nanosystems will be on the horizon, and nanobots will be considered a real possibility. The questions raised at this stage will include new diagnostic capabilities that lead to diagnoses of diseases long before cures are available, the extension of Moore's law beyond the limits predicted for current manufacturing strategies, and new sensors, just to name a few.

Midterm ethical issues will include: upheavals in global financial and manufacturing systems; marginalization of the poor (and perhaps also the rich, who are invested in older systems); new kinds of risk assessment; privacy and security issues raised by new capacities; implications for crime and the environment; new interest groups; a debate between optimists and pessimists about the prospects for these radical new technologies; and even questions about strong artificial intelligence and the status of computers (based on claims of people like Ray Kurzweil about "spiritual machines"). At this stage, there will be a mix of traditional and visionary questions, and coordination between government and industry will be necessary to address them.

In the long-term, Smith believes that although assemblers will still not be available, nanobots and "communicating and/or programmable nanosystems are becoming available," and a new kind of nanomedicine will emerge. On the basis of these capacities, many diseases will be cured, aging will be slowed, and a host of environmental and energy concerns will be solved. At this stage, some radical questions will be raised, such as what will happen if nanotechnology allows scarcity to become scarce; how much nanoprosthesis it will take to make a person nonhuman; how the concept of property will change if most things become replicable; if nanotechnology is as transformative as optimists predict, how difficult the transformation will be; what the implications will be of truly sentient artificial intelligences; how the nature of man will change; and how humans will/should interact with nanobots.

Smith believes these radical questions will arise fairly soon, whereas others (myself included) think they will not arise that quickly. Nevertheless, I think he rightly appreciates that the ethical questions about nanotechnology will "morph" as new capacities are introduced. Even the conservatives among us must acknowledge that by the end of the twenty-first century, many of these visionary scenarios will be realized, although they may come about in completely unexpected ways and have implications that cannot now be anticipated. In any case, it is not too soon to consider them seriously.

Smith's near-term nanotechnology closely resembles the loosely associated, more traditional kinds of research I addressed earlier under the rubric of the nanotechnology grab bag; his long-term nanotechnology approximates the more visionary definition (but without Drexler's universal assemblers, although Smith thinks they might be possible). In the short term, nanotechnology tends to be fragmented in a host of topical areas, with more global issues addressed in terms of future planning. As one moves from the short term to the long term, integration increases, as does the need for a more coordinated public/private response. A core task in anticipating these developments involves careful attention to the midterm vision, which should be formulated in a way that can guide the transition into an exciting, but disruptive, future. The midterm task of addressing ethical issues merges with the general task of characterizing nanotechnology; both jointly provide an anticipatory coherence of the emerging science and its interface with society.

NANOTECHNOLOGY AS AN ENABLING
SCIENCE AND TECHNOLOGY

In this essay, I have focused directly on nanoscale science and technology. However, some of the most significant ethical concerns are not about nano-technology itself, but about how this emerging science will converge with and make possible other radically new and developing technologies, especially those associated with biomedicine, information technology and robotics, and cognitive science. In the more radical visionary scenarios, this convergence leads to a "post-human" future (Joy, 2000; Kurzweil, 1999). The prospects and character of this convergence are beyond the scope of this presentation, but a discussion of the ethics of nanotechnology would not be complete without at least briefly address-ing issues in this area.

A major public/private initiative is under way to enhance human life by seeding the convergence of nanotechnology, biomedicine, information technol-ogy, and cognitive science (NBIC) (De Rosnay, 2001; Khushf, 2004b; Roco and Bainbridge, 2002). Leaders in NNI, such as Mike Roco, leaders of major corpo-rations, such as IBM and Hewlett Packard, and political leaders, such as Newt Gingrich and the current undersecretary of commerce for technology, Philip Bond, just to mention a few of the most prominent figures, are all involved in this initiative. The purpose is to establish a knowledge base and infrastructure to integrate current areas of rapidly developing science and direct efforts toward improving the human condition.

Radical changes are being contemplated. It is believed that in 10 to 20 years we could significantly alter the aging process, develop human-machine inter-faces, realize goals of space exploration, develop advanced robotics, and create smart environments. The summary NBIC document describes the longer term implications (Roco and Bainbridge, 2002):

> The twenty-first century could end in world peace, universal prosperity, and evolution to a higher level of compassion and accomplishment. It is hard to find the right metaphor to see a century into the future, but it may be that humanity would become like a single, distributed and interconnected "brain" based in new core pathways of society. This will be an enhancement to the productivity and independence of individuals, giving them greater opportunities to achieve personal goals.

The architects of this initiative appreciate that such radical prospects will require comprehensive discussions and debate to address ethical issues, which include broad human and environmental goals and potential impacts on every aspect of society. Core challenges are associated with balancing individual and communal well-being, a task that has always been central to ethics and social policy, as well as conceptualizing the notion of human flourishing that should guide the initiative. Although fairly specific goals have been put forth—namely, NBIC convergence for the purposes of human enhancement—the initiative could be understood in a more general way as a forum for exploring the future impact of all science and engineering, including qualitative changes just over the horizon.

Nanotechnology is the key enabling technology that will make possible NBIC convergence; thus serious reflection on these enabling capacities will require an approach that integrates issues raised by many other areas of science and engineering and issues raised by nanoscience and nanotechnology. The NBIC documents call for an increasingly integrated approach to sciences and technologies, and they suggest a conceptual framework for a holistic understanding. The development of this framework will require philosophical, ethical, and social analysis because it will surely influence how diverse activities associated with the research enterprise are integrated with each other and with the rest of society (Khushf, 2004b).

THE URGENCY OF THE TASK

I would like to close on a cautionary note. Although some radical developments, such as those associated with human-machine interfaces and smart environments, are already in the early stages of implementation (Maguire and McGee, 1999; Moore, 2003; Nicolelis, 2003; Nicolelis and Chapin, 2002; Roco and Bainbridge, 2002), I do not believe the qualitative difference between nanotechnology and other more conventional technologies will be apparent in the near term. Most research and commercialization is currently directed toward fairly traditional ends, and the first nanoproducts will be far from revolutionary. This does not mean we can take our time about reflecting on ethical issues. Unless we focus significant efforts on the ethical and social issues, the debate could be framed in a way that could make it extremely difficult to respond constructively to the radical capacities on the midterm horizon. I believe it is imperative that we put forth extensive efforts to address these ethical issues now.

There are already some indications of problems ahead. Consider, for example, the potential impact of the upcoming movie of Michael Crichton's book, *Prey*, in which swarms of self-replicating nanobots emerge as a threat to all of humanity. As one reviewer has said, "[p]ut Hollywood and Michael Crichton together and you've got the next big science scare" (Smith, 2003). Although the book is purely science fiction, Crichton begins with an introduction and ends with references that give the impression that his story is based on current science. The project described in the book resembles the Drexlerian project, and Crichton cites Drexler in the introduction; and the fictional evil company, Xymos, seems to be patterned after Zyvex. The book is a version of the dystopian scenario of Bill Joy (2000).

Many leaders in nanotechnology are greatly concerned that public perceptions will be formed by these images, which could lead to a reaction against the whole research enterprise of nanotechnology. Debates like the one between Drexler and Smalley on the feasibility of assemblers may become part of the public perception of nanotechnology, with science and movie images merging.

There are also indications that some current nanoproducts, such as nanorods, might have toxic effects, raising questions about the health and environmental impacts of nanotechnology (Smith, 2001; Wardak and Rejeski, 2003). It is possible that public debate on the toxic effects of current nanoproducts will resemble the debate about genetically modified organisms. Even worse, these discussions might become enmeshed in the assembler and dystopian debate, which could lead to a complex mixing of the meanings of nanotechnology.

It is difficult enough for researchers to tease out the diverse meanings of nanotechnology, and it may be impossible for the public. Thus, the public could become polarized, with some people advocating for and others advocating against the whole initiative. Examples of such polarization can be found in the debates on nuclear technology and genetically modified foods. But there is one significant difference. In the end, we cannot choose to forego nanotechnology. Nanoscale science and technology are too broad, and they signify developments in all of the sciences. Foregoing nanotechnology would be like foregoing chemistry or physics.

We must find a way to make some clear distinctions to frame the debate about nanoscale science and engineering activities (Roco and Bainbridge, 2001). And we should attempt to open this debate before polarization occurs. As Arnall (2002) notes, "both precautionary principle and industry advocates agree that there is time to create dialogue and consensus that could prevent the kind of confrontations . . . that plagued the development of biotechnology." To do that, we will have to develop the right kinds of visions, situated ethical theories, and topically based distinctions to guide the debate, and we will have to ensure that they are widely disseminated. Distinctions must be made in a way that anticipates and guides public debate.

Unfortunately, we are just beginning to address these issues in a nuanced way. In many ways, we are unprepared for a debate that is already at hand. To think through the challenges ahead, we will need the same kind of exponential growth in ethics research that is taking place in nanotechnology (Mnyusiwalla et al., 2003). In fact, we have a lot of catching up to do already.

A whole new kind of science and technology lies ahead, with capacity to alter humanity in unprecedented ways. We will need a new kind of dialogue to enable us to think through these capacities in a mature and responsible way.

REFERENCES

Alivisatos, A.P. 2001. Less is more in medicine. Scientific American 285(3): 67–73.

Arnall, A.H. 2003. Future Technologies, Today's Choices: Nanotechnology, Artificial Intelligence and Robotics: A Technical, Political and Institutional Map of Emerging Technologies. London: Greenpeace Environmental Trust.

Ashley, S. 2001. Nanobot construction crews. Scientific American 285(3): 84–85.

Baird, D. 2004. Thing Knowledge: A Philosophy of Scientific Instruments. Berkeley, Calif.: University of California Press.

Baird, D., and A. Shew. 2002. Probing the history of scanning tunneling microscopy. Available online at *www.cla.sc.edu/cpecs/nirt/papers.*

Baum, R., E. Drexler, and R. Smalley. 2003. Nanotechnology: Drexler and Smalley make the case for and against "molecular assemblers." Chemical and Engineering News 81(48): 37–42.

Beauchamp, T.L., and J.F. Childress. 2001. Principles of Biomedical Ethics, 5th ed. New York: Oxford University Press.

Buchand, G., and C.D. Montemagno. 2000. Constructing organic/inorganic MEMS devices powered by biomolecular motors. Journal of Biomedical Microdevices 2(3): 179–184.

Center for Responsible Nanotechnology. 2002. Current Results of Our Research. Available online at *http://crnano.org/administration.html.*

Crichton, M. 2002. Prey. New York: Avon Books.

De Rosnay, J. 2001. From molecular biology to biotics: the development of bio- , info- and nano-technologies. Cellular and Molecular Biology 47(8): 1261–1268.

Drexler, K.E. 1986. Engines of Creation: The Coming Era of Nanotechnology. New York: Anchor Books, Doubleday.

Drexler, K.E. 1992. Nanosystems: Molecular Machinery, Manufacturing, and Computation. New York: John Wiley & Sons.

Drexler, K.E. 2001. Machine-phase nanotechnology. Scientific American 285(3): 74–75.

Engelhardt, H.T. 1996. Foundations of Bioethics. New York: Oxford University Press.

ETC Group. 2003. The Big Down: From Genomes to Atoms. Winnipeg, Manitoba: ETC Group.

Etzkowitz, H. 2001. Nano-science and Society: Finding a Social Basis for Science Policy. Pp. 121–128 in Societal Implications of Nanoscience and Nanotechnology, edited by M. Roco and W. Bainbridge. Dordrecht, The Netherlands: Kluwer Academic Publishers.

Feynman, R. 1992. There's plenty of room at the bottom. Journal of Microelectromechanical Systems 1(1): 60–66.

Feynman, R. 1993. Infinitesimal machinery. Journal of Micromechanical Systems 2(1): 4–14.

Foresight Institute. 2000. Foresight guidelines on molecular nanotechnology, revised version 3.7. Available online at *www.foresight.org.*

Forest, D. 1989. Regulating nanotechnology development. Available online at *http://www.foresight. org/NanoRev/Forrest1989.html.*

Freitas, R.A. 2001. The gray goo problem. Available online at *KurzweilAI.net* March 20, 2001.

Fukuyama, F. 2002. Our Posthuman Future: Consequences of the Biotechnology Revolution. New York: Farrar, Straus and Giroux.

Galison, P.L. 1997. Image and Logic: A Material Culture of Microphysics. Chicago: University of Chicago Press.

Galleo, K. 2003. Bottom-up, top-down or self-assembly required. While we're sitting here in Drexler's waiting room, should it be up to chemists to define nanotechnology? Smalltimes 3(5): 8.

Goldstein, A.H. 2003a. Invasion of the high-tech body snatchers. Available online at *http://www. salon.com/tech/feature/2003/09/30/bioengineering/print.html.*

Goldstein, A.H. 2003b. Nature versus nanotechnology: reinventing our world one atom at a time. Available online at *http://www.shelleconomistprize.com/essays/3rd_prize_Alan_Goldstein.pdf.*

Gorman, M.E. 2002. Combining the Social and the Nanotechnology: A Model for Converging Technologies. Pp. 325–330 in Converging Technologies for Improving Human Performance, edited by M. Roco and W. Bainbridge. Arlington, Va.: World Technology Evaluation Center, Inc.

Joy, B. 2000. Why the future doesn't need us. Wired 8(4): 238–262.

Keiper, A. 2003. The nanotechnology revolution. The New Atlantis: A Journal of Technology and Society 1(2): 17–34.

Khushf, G., ed. 2004a. Handbook of Bioethics: Taking Stock of the Field from a Philosophical Perspective. Dordrecht, The Netherlands: Kluwer Academic Publishers.

Khushf, G. 2004b. Systems theory and the ethics of human enhancement: a framework for NBIC convergence. Annals of the New York Academy of Sciences 1013.

Kurzweil, R. 1999. The Age of Spiritual Machines. New York: Penguin Books. Related links can be found on *www.KurzweilAI.net.*

Lieber, C.M. 2001. The incredible shrinking circuit. Scientific American 285(3): 59–64.

Maguire, G.Q., and E.M. McGee. 1999. Implantable brain chips?: time for debate. Hastings Center Report 29(1): 7–13.

McKibben, B. 2003. Enough: Staying Human in an Engineered Age. New York: Times Books, Henry Holt and Company.

Mickelson, E.T. 2003. . . . and neither will the government. Smalltimes 3(1): 8.

Mnyusiwalla, A., A.S. Daar, and P.A. Singer. 2003. "Mind the gap": science and ethics in nanotechnology. Nanotechnology 14: R9–R13.

Moore, M.M. 2003. Frontiers of Human-Computer Interaction: Direct-Brain Interfaces. Pp. 47–51 in Frontiers of Engineering: Reports on Leading-Edge Engineering from the 2002 NAE Symposium on Frontiers of Engineering. Washington, D.C.: The National Academies Press.

National Commission for the Protection of Human Subjects. 1978. The Belmont Report: Ethical Guidelines for the Protection of Human Subjects Research. Washington, D.C.: U.S. Department of Health, Education and Welfare.

Nicolelis, M.A.L. 2003. Brain-machine interfaces to restore motor function and probe neural circuits. Nature Reviews/Neuroscience 4: 417–422.

Nicolelis, M.A.L., and J.K. Chapin. 2002. Controlling robots with the mind. Scientific American 287(4): 46–53.

NRC (National Research Council). 2002. Small Wonders, Endless Frontiers: A Review of the National Nanotechnology Initiative. Washington, D.C.: National Academy Press.

NSTC (National Science and Technology Council). 1999. Nanotechnology: Shaping the World Atom by Atom. Washington, D.C.: National Science and Technology Council.

NSTC. 2002. National Nanotechnology Initiative: The Initiative and Its Implementation Plan. Washington, D.C.: National Science and Technology Council.

Phoenix, C., and M. Treder. 2003. Three Systems of Action: A Proposed Application for Effective Administration of Molecular Nanotechnology. Center for Responsible Technology. Available online at *http://crnano.org/systems.htm.*

Reynolds, G.H. 2001. Environmental regulation of nanotechnology: some preliminary observations. Environmental Law Reporter 31: 10685.

Reynolds, G.H. 2002. Forward to the Future: Nanotechnology and Regulatory Policy. San Francisco: Pacific Research Institute.

Roco, M.C. 2003a. Broader societal issues of nanotechnology. Journal of Nanoparticle Research 5: 181–189.

Roco, M.C. 2003b. Nanotechnology: convergence with modern biology and medicine. Current Opinion in Biotechnology 14: 337–346.

Roco, M.C., and W.S. Bainbridge, eds. 2001. Societal Implications of Nanoscience and Nanotechnology. Dordrecht, The Netherlands: Kluwer Academic Publishers.

Roco, M.C., and W.S. Bainbridge, eds. 2002. Converging Technologies for Improving Human Performance: Nanotechnology, Biotechnology, Information Technology and Cognitive Science. NSF/DOC-Sponsored Report. Arlington, Va.: World Technology Evaluation Center, Inc.

Roco, M.C., R.S. Williams, and P. Alivisatos, eds. 1999. Nanotechnology Research Directions: IWGN Workshop Report/ Vision for Nanotechnology R&D in the Next Decade. Dordrecht, The Netherlands: Kluwer Academic Publishers.

Roukes, M. 2001. Plenty of room indeed. Scientific American 285(3): 48–57.

Roy, R. 2002. Giga science and society. Materials Today 5(12): 72. Also available online at *http:// www.materialstoday.com/pdfs_5_12/opinion.pdf.*

Smalley, R.E. 2001. Of chemistry, love and nanobots. Scientific American 285(3): 76–77.

Smalltimes. 2003. Lending Life a Hand: Special Life Science Issue. Smalltimes 3:4.

Smith, D. 2003. Brave New Tiny World of the Nano. Sydney Morning Herald, November 15: 1. Also available online at *http://www.smh.com.au/articles/2003/11/14/1068674381005.html.*

Smith, R.H. 2001. Social, Ethical, and Legal Implications of Nanotechnology. Pp. 257–271 in Societal Implications of Nanoscience and Nanotechnology, edited by M. Roco and W. Bainbridge. Dordrecht, The Netherlands: Kluwer Academic Publishers.

Stix, G. 2001. Little big science. Scientific American 285(3): 32–37.

Stuart, C. 2003a. Environmental applications help in cleanup efforts, manufacturing. Smalltimes 3(1): 40–43.

Stuart, C. 2003b. Nano's balancing act: remarkable rewards are weighed against possible risks. Smalltimes 3(1): 34–36, 38–39, 44.

Theis, T.N. 2001. Information Technology Based on a Mature Nanotechnology: Some Societal Implications. Pp. 74–84 in Societal Implications of Nanoscience and Nanotechnology, edited by M. Roco and W. Bainbridge. Dordrecht, The Netherlands: Kluwer Academic Publishers.

Venneri, S.L. 2001. Implications of Nanotechnology for Space Exploration. Pp. 213–218 in Societal Implications of Nanoscience and Nanotechnology, edited by M. Roco and W. Bainbridge. Dordrecht, The Netherlands: Kluwer Academic Publishers.

Vigliotti, A. 2003. A no-nonsense nano definition. Smalltimes 3(7): 4.

Wardak, A., and D. Rejeski. 2003. Nanotechnology and Regulation: A Case Study Using the Toxic Substance Control Act (TSCA). Publication 2003-6. Washington, D.C.: Woodrow Wilson International Center for Scholars.

Weil, V. 2001. Ethical Issues in Nanotechnology. Pp. 244–251 in Societal Implications of Nanoscience and Nanotechnology, edited by M. Roco and W. Bainbridge. Dordrecht, The Netherlands: Kluwer Academic Publishers.

Weingart, P. 2002. The moment of truth for science. European Molecular Biology Organization (EMBO) Reports 3(8): 703–706.

Whitesides, G.M. 2001. The once and future nanomachine. Scientific American 285(3): 78–83.

Whitesides, G.M. 2002. Beyond molecules: self-assembly of mesoscopic and macroscopic components. Proceedings of the National Academy of Sciences 99(8): 4769–4774.

Whitesides, G.M., and J.C. Love. 2001. The art of building small. Scientific American 285(3): 39–47.

Neurotechnology and Brain-Computer Interfaces
ETHICAL AND SOCIAL IMPLICATIONS

PAUL ROOT WOLPE
Center for Bioethics
University of Pennsylvania

I want to thank my friend George Khushf, who without knowing it, perfectly set up what I want to talk to you about. All of the things he mentioned at the end of his talk—a retinal prosthesis, Miguel Nicolelis' monkey and robotic arm, and "roborats"—are things I'm going to discuss.

Neuroethics is a brand new field. The modern use of the term was coined by William Safire, of all people, in *The New York Times*, who is on the board of the Dana Foundation and is very interested in issues of the brain. Neuroethics is a field that looks at emerging technologies and their relation to the brain. In Europe, the term has been used to refer to the clinical care of people with strokes and other neuropathologies. In the United States, it has come to mean something different. Here is the technical definition, which I wrote for the *Encyclopedia of Bioethics*: "The field of neuroethics involves the analysis of, and remedial recommendations for, the ethical challenges posed by chemical, organic, and electro-mechanical interventions in the brain" (Wolpe, 2004.).

Neuroethics includes, for example, the proper use of psychopharmacology, which is of course a long-standing issue. Human beings have been trying to enhance the brain with chemicals ever since they discovered fermentation, perhaps even before that. And we are still doing it. In fact, we have gotten a lot better at it, a lot more specific about it. We have created effective, highly specific drugs that can alter moods and cognitive states in very selective ways. We have become a psychopharmacological culture. As soon as new psychotropic drugs or other designer drugs come out, we use them, and even as we are using them, we wring our hands over whether we should. But our concerns don't seem to keep us from buying them, whether the drug is Prozac, Ritalin, Viagra, or another drug, or even

57

nonpharmaceutical enhancements. We are long past using these drugs solely for identified pathologies; we use them now to micromanage our moods and cognitive states.

I am a sociologist by training, and an underlying given of sociology is that all of science occurs within a cultural context. That is, scientists and engineers ask questions that they then try to solve. Where do the questions come from? People in different cultures ask different questions, and people in different historical periods ask different questions. In fact, very often in the history of science, theories disappear, not because they have been disproved, but because society is no longer interested in them. The questions we ask of science change as societies evolve. But it is crucial that we understand that the questions themselves have embedded values and ethical components.

A perfect example is the ecological movement. Solving the problems of ecology through science was considered a silly idea by a lot of people 30 or 40 years ago. People then just didn't think in those terms. Now, of course, ecological concerns are ubiquitous, and the idea that science can provide solutions to those questions is very much in everyone's consciousness.

So, the questions we ask of ourselves, and not just the answers we give, have profound ethical implications. What problems are we really trying to solve? And why have we chosen these problems and not other problems? Why, for example, do we put such a premium on enhancing cognitive function? Why do we think of ourselves as mechanisms that can be improved? These questions can be traced to the history of our particular time and place. The questions are different in different countries—as George said, Europeans think about these issues differently from the way we think about them. Compare our attitude toward genetically modified foods, for example, with the Europeans' attitude.

Another example is how we describe the human body. We think in terms of genetics and metaphors of information technology. Bill Wulf was talking earlier about how intervening in a single bit of a complex software code can cause problems; well, so, my Microsoft Office crashes if there is an error. A single-bit alteration in the genetic code, however, can cause cancer or some other genetic disease. When we talk about small interventions in the 3 billion bits (as opposed to 10 to 200 zeros) that make up the blueprint of my cells, I'm very concerned about a single-bit mistake. At the National Aeronautics and Space Administration (NASA), where I am chief of bioethics, we are debating the issue of radiation exposure. What is an appropriate amount of time to allow a mutagenic force to impact the bodies of our astronauts? How much alteration in an astronaut's genetic code is allowable? How many "bugs" are acceptable in the human software? Note the convergence of metaphors. We can talk about computer code and the human body seamlessly because the questions and language of this moment in time are the same whether we are talking about biology or computers.

When we talk about psychotropic drugs, we use similar metaphors—the brain as computer, a neuron as a single switch, the brain as wetware containing software. But fundamental to neuroethics, and the engineering of the wetware between our ears, is the question of what it means when we begin to intervene in that software, in the functioning of our brains. Intervening in the functioning of a computer is very different from intervening in the functioning of our genes or our neurons.

We also think about it differently because we are cerebrocentric. If we took my brain and put it in George Khushf's body, aside from my becoming slightly more handsome, you would still think it was me, not George. George's body would become the receptacle for me, because we believe our personalities and everything important about us, or at least most of what's important about us, resides in our brains. That, however, is a culturally and historically specific claim. The site of personhood to the Japanese is in the gut, which is one reason the Japanese have been resistant to the idea of brain death and transplants. Their resistance is not based on a Luddite resistance to technology—the Japanese love technology. The brain death criteria violate their cultural model of where personhood resides.

And so, when we talk about psychotropic drugs, when we "listen to Prozac," as Peter Kramer says, when someone says "the real me was evoked by taking a psychotropic drug, and I was never me until I took Prozac," we have to ask what "me" means in that sense. What is the nature of our sense of identity? What is mind if we can alter it in profound ways? These are profound cultural questions.

When we use drugs to alter mental processes in children, the questions become even more profound. Ritalin prescription patterns, for example, are bimodally distributed. Ritalin use is very high in wealthy suburban schools and in poor inner city schools. In wealthy suburban schools, the main drivers behind Ritalin use are parents who want their kids to have the extra edge a good amphetamine can give them. In inner city schools, the people who drive Ritalin use are school managers who use it as a tool to manage problem kids. And so, it's not just the existence of the drug, but also how we use it, and under what circumstances, that can have profound implications. We also give kids antidepressants. A lot of the pediatric literature says this is a great thing, that depressed kids have been undertreated. Everyone agrees that there is depression in children. But many of these drugs have never been tested on children because it is expensive and difficult to do clinical trials on children and because drug companies know that, even if they don't test them on kids, doctors will prescribe them because they have no choice. So the drug companies get the income without making the investment, and we put ourselves into a Catch-22 situation. We don't like to test things on children, so we give them without ever testing them.

And finally, one of the most profound questions in pediatrics in the future may be about prophylactic treatment. Once we become skilled at understanding brain imaging and the morphological features of the brain, we will be able to predict psychiatric susceptibility. We will be able to identify prodromal states in certain diseases. We may be able to image a child and say, "This brain looks like the kind of brain we see in schizophrenics, or it looks like susceptibility is very high because of morphological or functional features that we see through PET or functional MRI." Should we treat the child prophylactically? Will we have scores of children on drugs who show no symptoms whatsoever but who seem to have pre-pathological brains? That is going to be an important question in the future that pediatrics has barely begun to address.

The issue on the horizon is the use of psychotropic drugs as lifestyle drugs, which will force us to confront questions about the nature of personality, selfhood, and human enhancement. Very soon, we are all going to be micromanaging our moods. We are going to replace our current liquid caffeine delivery systems with wake-up pills to get us up in the morning and get us dressed and ready to greet the day. Right before we get to work, we will take a get-ready-for-work pill that focuses our attention. Right before lunch, we'll take a pill that mellows us out for an hour, and also probably a pill to prevent our bodies from absorbing the fat and carbohydrates we're about to eat at lunch. After lunch, we'll take a pill that makes it so we don't have the post-lunch depression we've all experienced (that's why I prefer to speak before lunch rather than right after lunch at these meetings). When we get home, we'll take "Sublime," a pill that puts us in the mood to see our families again.

This is already happening. Here is an advertisement from menshealthworld. com. "Consult with your doctor." For what? The ad tells you below: "Celebrex, Propecia, Viagra, and Xeneca," a weight loss drug. Pretty soon physicians will become lifestyle pharmacists. Actually, physicians will become irrelevant for that purpose because you can already get most of these drugs on the Web. A recent study of websites showed that you can get Viagra everywhere, simply by answering a few questions. We all get the spam, right?

But you should all go to one of those websites and place an order—you don't have to put in your Visa card number, but go as far as you can until you chicken out on the process. You will find that there is a clinical input form you must fill out that is "checked by a physician" to make sure you qualify for Viagra. There are four questions. If you get them wrong, you can try again until you get them right.

Then there is brain imaging, which is bringing up a whole series of new, interesting questions. It turns out that the phrenologists were right, at least to a degree. Some morphological features of the brain actually do correlate with personality traits. For example, some studies have shown that you can predict some things about people, like whether they are socially withdrawn or socially active, by the size of the cingulate gyrus. Just by size, it correlates well.

And there are features of the brain that can be identified by their character. For example, a history of major depression or significant drug use (cocaine or other drugs) can leave morphological signs on the brain that can be seen on CAT scans, although not individually yet. If you look at a group of people who are now in total remission or who don't use cocaine but were once cocaine addicts, and you compare them to "normal" people, you can see statistical differences in their brain structures. At this point, though, we can't say that a single individual was a cocaine addict, except at the extremes.

But imagine what would happen if we started using brain imaging routinely in the public sector—I'll explain why we would in a moment. But imagine what that would mean for privacy. All kinds of traits might be revealed that we might not want known. Drug abuse is not the only thing that can be seen. For example, the best way to tell if someone is a drug addict is to expose them to the drug—a picture of it or the smell of it. You can see excitation in the brain, even if they no longer use the drug. You can also do that with sex offenders or people who aren't sex offenders but who have a sexual proclivity. If you want to know if someone has a particular fetish, expose them to the fetish, and look at their functional MRI.

We are talking about enormous possibilities for invasion of privacy here. We are talking about the ability to use technologies for social screening. Believe me, NASA would like nothing better than to put astronauts into brain scanners and say, "Looks like we have a pilot here—great visual cortex, good spatial sense." That is a pipe dream, of course, but NASA is looking for any piece of information that might improve their chances. Aptitude tests might soon be replaced by brain scans.

Here's another example. We are very bad at detecting lying. However, Daniel Langleben at the University of Pennsylvania recently did a study in which he put people into an fMRI scanner, gave them a card, and told them to lie at some point about which card they had. Through brain imaging, he found he could actually detect the difference in grouped data between lies and truthfulness.

Yet there are problems with such studies. There is a big difference between looking at a card with a cross on it and saying "star," and telling a lie, like "I did not have sex with that woman, Monica Lewinsky." These are very different kinds of lies. And it is unlikely that brain scanning can detect a lie as complex and robust as the latter, if we want to call that a lie. It depends, of course, what the meaning of "sex" is. But these are fascinating questions.

Here is another interesting study. The amygdala is a part of the brain that plays a role in emotions, such as fear. In one study, white males with high racism scores were put into PET scans. When they were shown pictures of white faces, there was very little response. Famous black faces that they recognized, like Martin Luther King, elicited no response. But when they were shown unfamiliar black male faces, their amygdalas lit up like Christmas trees. Evidence of racism in a brain scan!

There is talk now about creating remote brain scanners for airports. If you walk through an airport and your amygdala is lit up, it may mean you are a terrorist or it may mean you had a fight with your wife or it may mean you're a white racist. In any case, you would be brought to the back room to be strip searched.

Brain imaging presents incredibly difficult problems that must be solved before we can use it in social settings—to screen for employment, for honesty, for sensitive jobs; to detect lies; to track kids into aptitudes; and so on. These are all things that the Defense Advanced Research Projects Administration (DARPA) and other defense agencies are very interested in.

Let's move for a moment to regenerative neurology. We can put actual fetal nigral cells into people's brains for Parkinson's and other diseases. In the United States, the ethical conversation about that issue was focused entirely on where we would get the fetal cells; abortion was central to our discussion. In Europe, they asked a different question—what it would mean to put cells into someone's brain from someone else's brain? Would you adulterate their personhood? And for research in Scandinavia they decided that, yes, you could put cells into someone's brain, fetal nigral cells, but they have to be disaggregated. You cannot put a clump of brain that might have its own coherent integration into someone else's brain. We never even had that conversation here in the United States.

I also want to say a quick word about deep-brain stimulators. I was in an operating room last week to observe neurosurgeons putting a deep-brain stimulator into the putamen of a person with Parkinson's. It was an amazing thing to see. They threaded the device through a canula deep in the brain. You watched the progression of this thing, by tenths of a millimeter, deep into this person's brain, and you listened to the brain activity through an audio feed at the end of the probe. When they got into the putamen, the static-y sound of neurons firing increased significantly. You could just hear the cells firing wildly. The patient was on the table shaking, and when they turned the thing on, his tremors all ceased, in a moment, and he cried out, "Ah!" This man hadn't been able to feed himself in years. They handed him a glass and said "pretend you are drinking beer," and he brought the cup smoothly to his mouth. When they turned the current off to thread the wire through the inside of the skin to the stimulator implanted under the clavicle, he cried out again, "No, don't turn it off!" They assured him it would go back on, but it was an amazing thing to see.

There are some reports now, very preliminary, that the spouses of people who have gotten these deep-brain stimulators are reporting that, you know, "George doesn't seem exactly like George anymore." Is that because they have gotten used to the George who has had Parkinson's for five years? Or could it be that deep-brain stimulators, even though they are in motor centers of the brain, are actually evoking some kind of change? Nobody knows.

Finally, I want to talk about brain-computer interfaces. First noninvasive interfaces. A number of EEG-based technologies use action potentials to translate brain impulses into action. The problem is that the skull is a very bad conductor, a very bad transmitter of the electrical activity of the brain. So, when you put these things in and these caps on, you muffle most of the activity you want to detect. Using the P300 evoked-response potential, you can tell when the brain acts as an entire brain, going "Ah-ha, that's it!" Based on that idea, they have now created computer-brain interfaces that allow people to move cursors around screens and all kinds of things without any implanted technologies.

A similar technology is brain fingerprinting. Lawrence Farwell has used this technology forensically. The concept is to use EEGs to show whether a person is looking at a familiar or unfamiliar scene. A suspect is shown rapid pictures, and then, boom, a picture of a crime scene where, for example, the suspect says he never was. If his brain shows familiarity, Farwell can say he was probably there. When a suspect says he was not there, and the prosecutor claims he was there, if the suspect's brain shows unfamiliarity, Farwell can say with even more confidence that he probably wasn't there. This technology was ruled admissible in the Terry Harrington case, and he was let go. This brings up a whole range of important issues in jurisprudence. Believe me, if the Bush administration could get its hands on some of these technologies, they would be on a plane to Guantanamo Bay tomorrow.

Some brain-computer interfaces are implantable, rather than transcranial. These include cochlear implants and the optic nerve implant. Researchers are also working on retinal prosthetics. Today they have about 16 electrodes. A prosthesis with 1,000 or so electrodes could allow a patient to really look at things, to read a book, for example. But in the meantime, a person who is stone blind can read the top two lines of an eye chart. This is a fascinating prosthetic possibility.

Drs. Bakay and Kennedy of Emory University have a patient named Johnny Ray, JR they call him, who had a brain-stem stroke. JR is completely locked in, completely paralyzed, can't communicate in any way. Electrodes were implanted in his brain, and he was taught to move a cursor around a board to point to phrases, such as "I'm uncomfortable," "thanks for visiting," and so on. Now he has begun to use an alphabet to spell out his name.

Or take Miguel Nicolelis who is in the news because of a paper that was just released about a new technology. Nicolelis has put electrodes in the brains of owl monkeys, 30 or 40 electrodes in one, 200 in another, and then had them remotely control robotic arms. Nicolelis and his team determined what the monkeys' brain waves looked like when they moved their own arms, then used algorithms to translate them and taught the monkeys to control robotic arms. As the monkeys realized that the robotic arms mimicked the movement of their own arms, they eventually dropped their arms and began to control the robot arms entirely with their brains, without moving their arms. Amazing!

And then there is the "roborat," a rat with electrodes controlling its movements. The roborat has been turned, basically, into an organic robot. It no longer has the ability to make decisions about where it wants to go. The animal's behavior is governed by electrodes activated by someone using a joy stick who determines whether the animal moves right or left or goes up or down a tree. This kind of technology raises all kinds of ethical questions about using technology to control animals.

A man named Kevin Warwick had a chip implanted in his arm, which he then connected to the computer and environment in his laboratory. When he enters the lab, the lights go on and jazz starts playing. His heart beat and blood pressure appear on his computer screen. Warwick said that he suddenly began to feel connected to the environment in a way he hadn't before.

The common feature of these technologies is that they use technologies to control moods, cognitive functions, and physical functions. Through these technologies, we can begin not only to enhance ourselves, but also to connect ourselves to our environment in new ways. It is already happening. *Science* recently printed a cover article about bionic humans. In other words, we are becoming cyborgs, not in the science fiction sense, but in a practical, real, obvious sense; our technology will be integrated into our bodies, and our bodies will be integrated into our technology in a seamless way. This may not turn us into "spiritual machines," as Ray Kurtzweil claims, but it will certainly turn us into spiritual man-machine hybrids. The ethical question that confronts us is: who will have control of these technologies, and who will determine their ethical nature? Who will protect our privacy? Who will ask the important questions about enhancement—when it is good, when it is bad, and who should or should not have it? Or will these products be put on the consumer market for consumer response?

Many people are already trying to answer these questions in the negative. For example, Bill McKibben in his book, *Enough*; Leon Kass, the head of Bush's Presidential Bioethics Council, who has come out against in vitro fertilization, stem cell research, and other technologies; and Francis Fukuyama, who wonders in his book, *Our Posthuman Future*, if we are threatening "human nature." These and others are forces arrayed against these technologies.

Others are advocates for these technologies. Just as there are nanophobes and nanophiles, there are neurophobes and neurophiles defining the arguments. Science cannot march too far ahead of ethics, not because as an ethicist I need to be employed—that's always a good thing—but because ethics is going to determine how neurotechnologies are received by the public.

We made a mistake with the cloned sheep, Dolly, which was presented to the public without prophylactic ethical conversation. The public response was international hyperventilation because people didn't understand what Dolly meant, what the implications were. We must engage the public in this conversation before these technologies are developed further. We all have a stake in the outcome.

REFERENCES

Fukuyama, F. 2002. Our Posthuman Future: Consequences of the Biotechnology Revolution. New York: Farrar, Straus, and Giroux.

Kramer, P.D. 1993. Listening to Prozac: A Psychiatrist Explores Antidepressant Drugs and the Remaking of the Self. New York: Viking.

McKibben, W. 2003. Enough: Staying Human in an Engineered Age. New York: Henry Holt.
Wolpe, P.R. 2003. Neuroethics. Pp. 1894–1898 in Encyclopedia of Bioethics, vol. 4, 3rd ed., edited by Stephen G. Post. New York: Macmillan Reference USA.

E³

ENERGY, ENGINEERING, AND ETHICS

JOHN F. AHEARNE
Sigma Xi, The Scientific Research Society
Research Triangle Park, North Carolina

The President's Committee of Advisors on Science and Technology said this about the importance of energy (PCAST, 1997):

> The United States faces major energy-related challenges as it enters the twenty-first century. Our economic well-being depends on reliable, affordable supplies of energy. Our environmental well-being—from improved urban air quality to abating the risk of global warming—requires a mix of energy sources that emits less carbon dioxide and other pollutants than today's mix does. Our national security requires secure supplies of oil or alternatives to it, as well as of prevention of nuclear proliferation. And for reasons of economy, environment, security, and stature as a world power alike, the United States must maintain its leadership in the science and technology of energy supply and use.

Economically, expenditures on energy account for 7 to 8 percent of gross economic product in the United States and worldwide and a similar fraction of the value of U.S. and world trade. Furthermore, environmentally, energy supply accounts for a large share of the most worrisome environmental problems at every geographic scale.

Outstanding issues include: how energy is obtained (from which countries, from which fuels, and in what way); how it is used (in what quantity, which brings in conservation, energy efficiency, the energy portfolio [hydro, solar, wind, geothermal, nuclear, natural gas, coal]); and how it is distributed and transferred. All of these issues have ethical dimensions.

HOW ENERGY IS OBTAINED

The way we obtain energy is the subject of political battles. Should we open the Alaskan National Wildlife Refuge (ANWR) to oil exploration? Should we allow offshore drilling (e.g., off of Santa Barbara)? Should we open for drilling other lands controlled by the U.S. Department of the Interior (e.g., in the Rocky Mountains)?

There also are less obvious political battles, for example, means of support for certain kinds of energy. What kind of subsidies should be given, if any? Today, wind power enjoys a 1.7 cents/kwhr[1] production subsidy, which extends for 10 years (Deutch et al., 2003). The argument for subsidizing wind energy is that it is an emission-free, unlimited resource. But nuclear energy is also emission free, nearly unlimited, and, like wind power, expensive. Should nuclear energy also get a production credit? A recent MIT study concluded that nuclear energy should get the same credit as wind power (Deutch et al., 2003). Yet nuclear power is not included in the emission-free portfolio in the Kyoto agreements. Nuclear power in the United States does have the limited, largely misunderstood, Price-Anderson protection.

The United States gets its energy from many sources, but only a few are domestic. These include oil, coal, gas, nuclear power, and hydropower.[2] Every energy source has opponents.

Opposition to hydropower comes mainly from people who would like to restore the canyons that were flooded to create the reservoirs for power generation dams (e.g., the Hetch Hetchy reservoir) and people who want to remove dams to restore habitats (e.g., to allow fish, such as salmon, to return unimpeded to their spawning grounds). Opposition to natural gas comes primarily from people who object to new pipelines running through their areas. Although the combustion of natural gas, a fossil fuel, produces greenhouse gas, this objection is seldom raised because gas contributes only about half as much greenhouse gas as coal per unit of energy produced. Opposition to oil is focused on U.S. reliance on foreign oil and proposals to drill in protected areas. Nuclear power generates intense feelings both for and against, and objective analysis is not a trademark of the extremists on either side. (Many years ago, Daniel F. Ford of the Union of Concerned Scientists called nuclear power a religion in search of a bible.) Opposition to nuclear power is based on many factors: radioactive waste; fear of catastrophic accidents; the risk of proliferation; and the connection of nuclear power with organizations considered to be untrustworthy. For example, the rhetoric used in a recent presentation by Dr. Thomas B. Cochran (director of the

---

[1]Scott Kirsen cites 1.8 cents/kwhr in an article in *the New York Times*, "Wind Power's New Current," (August 28, 2003).

[2]Conservation and efficiency can all be treated as energy sources because they reduce the demand for energy from other sources.

nuclear program for the Natural Resources Defense Council], on September 3, 2003, at a National Research Council Board of Radioactive Waste Management meeting is illustrative. The U.S. Department of Energy (DOE) had proposed redefining some waste that was previously considered high-level waste as "incidental waste" and, therefore, able to be treated as low-level waste. The presenter had filed a suit in opposition to DOE in a federal court. The following examples from notes submitted to the board give an idea of the depth of feeling among those opposed to this action:

> DOE unlawfully closed two high-level waste (HLW) tanks...
> DOE unlawfully promulgated DOE Order 435.1...
> DOE used junk science to falsely portray this "incidental waste" as low-level waste.
> The first declaration of [the DOE witness] falsely states...
> [The DOE witness] attempted to mislead the Court...
> [The DOE secretary], in his...letter to Speaker of the House Dennis Hastert, repeats the baseless assertion...
> The Secretary misinforms Congress...
> [The DOE secretary] misinforms Congress when he implies ...
> This is simply false....

In the last quarter century, many ideological battles have been fought over federal funding for energy research and development (R&D). During the Nixon years, funding for nuclear power, particularly for breeder reactors, rose, but funding for solar and other renewable energy sources was stagnant. Under President Carter, solar and other renewables were favored, and Carter waged a major, but unsuccessful, battle to kill the breeder reactor program. But he did continue the policy announced at the end of the Ford administration not to support the reprocessing of nuclear fuel. The Reagan administration reversed course—being more in favor of nuclear power and less in favor of renewables. Reagan also reversed the policy on reprocessing, although to little effect because the economics were not favorable to reprocessing. Things changed again under Clinton, who was a strong supporter of renewables, at least in words, but who eventually zeroed research on nuclear power. The Bush administration has taken nuclear power out of the woodshed but has also supported renewables, perhaps in response to pressure by Congress.

HOW ENERGY IS USED

Energy is primarily used to heat, to cool, and to transport people and things. Here again problems have arisen. As recent blackouts in the United States and Italy have demonstrated, the electrical transmission and distribution system is fragile in both technical and human terms. Part of the solution may be to install more and higher voltage transmission lines. But siting these lines is extraordinarily difficult. (This is one reason the energy bill currently before Congress includes a controversial provision to allow a federal preemption for siting lines.) American Electric Power recently noted that it takes about 10 years to get approval for a relatively short high-voltage line. Sometimes it takes much longer.

Even highly touted wind power has run into problems. A proposed wind farm off Nantucket Island in Massachusetts has aroused vigorous opposition among local and summer residents. Coal plants are perceived to be "dirty"; opponents usually argue in favor of natural gas generation. But some utilities are wary of building natural gas plants, because generation costs would then be at the mercy of price rises in a single source.

Human problems were integrally involved in the Three Mile Island accident, the Chernobyl accident, the Japanese Toka Mura accident, the large hole in the reactor lid at the Davis-Besse plant, and the enforced closing of the 17-reactor fleet of the Japanese utility, TEPCO.

ETHICAL ISSUES

Many ethical issues are related to maintaining objectivity, a form of honesty. Unfortunately, ideology often trumps objectivity, and too often "the end justifies the means." I cannot believe some of the ardent supporters and ardent opponents of various energy sources really believe the harsh rhetoric they use. Many who espouse these positions put their trust in their leaders and do not think through the issues and arguments carefully. These trusted leaders are often very smart, and, I believe, they know that some of their statements are, at best, exaggerations. They are like some managers of federal programs who see only one side of an issue and who believe that nothing will go wrong and that things will turn out positively. Is this unethical? I believe it is.

Technology professionals have a responsibility to analyze issues rigorously and with complete objectivity. Many citizens do not have the background, the resources, the time, or the interest to dig deeply into technical issues. Therefore, the public must rely on the professionals, who, therefore, carry a heavy burden. The following examples show the importance of objective analysis by engineers.

We are familiar with the warnings about the O-rings that were not heeded and led to the destruction of the shuttle *Challenger*. Now we are hearing about similar unheeded warnings about the tiles on *Columbia*. On September 26, 2003, *The New York Times* carried a story about Rodney Rocha, the chief engineer in the Structural Engineering Division, Johnson Space Center. Five days into the *Columbia* flight, he and his coworkers reviewed pictures of foam breaking off and striking the left wing of the shuttle. Because they could not tell exactly where the foam hit, they thought an effort should be made to examine the wing. Mr. Rocha proceeded to send messages to management urging that satellite imagery be used. In one message, he wrote to upper level managers asking, "Can we petition (beg) for outside agency assistance?" His requests were denied, and one manager told him, "I'm not going to be Chicken Little about this" (Glanz and Schwartz, 2003).

Furious debates are under way about the Yucca Mountain repository. The majority of the technical community supports geologic repositories in general and Yucca Mountain in particular. Nevada has mounted a furious, years-long opposition. Given the way Yucca Mountain was chosen (by Congress), this opposition is rational; the selection process replaced the balanced process set up in the Nuclear Waste Policy Act of 1982 and violated the deliberative process recommended in several National Academies reports (NRC, 1989, 1996, 2001). However, independent groups, such as the Nuclear Waste Technical Review Board and the National Academies, have not found anything seriously wrong with the Yucca Mountain site (NRC, 1992, 1995, 2001).

Heated arguments have also arisen over the Environmental Protection Agency's (EPA's) new rule on new source reviews. The acting EPA administrator, Marianne Horinko, said in a statement, "The changes we are making in this rule will provide industrial facilities and power plants with the regulatory certainty they need. This rule will result in safer, more efficient operation of these facilities, and, in the case of power plants, more reliable operations that are environmentally sound and provide more affordable energy" (Energy Daily, 2003). A supporting comment was made by Thomas Kuhn, president of the Edison Electric Institute: "With the issuance of the final rule today, we are returning to the common-sense standard that has applied throughout most of the history of the Clean Air Act. Today's regulations will lift a major cloud of uncertainty, boosting our efforts to provide affordable, reliable electric service and clean air."

However, S. William Becker, executive director of the State and Territorial Air Pollution Program Administrators and the Association of Local Air Pollution Control Officials, said, "This rule eviscerates the NSPR [New Source Performance Review] program and represents a huge step backward in our efforts to achieve and sustain clean air. Not only will it degrade existing protections of public health and environment, it will be very difficult to implement and enforce" (Energy Daily, 2003). Rebecca Stanfield, staff attorney for United States Public Interest Research Group, also weighed in on the subject: "For decades to come, Americans will be forced to breathe air containing more harmful smog and soot because of the action the Bush administration is taking today" (Energy Daily, 2003).

Other programs in the offing are raising issues related to engineering accuracy. For example, uninformed enthusiasm is growing among lawmakers and the public for hydrogen as a magic fuel and for essentially inexhaustible energy from fusion. Both of these will be extremely hard to develop. Engineers understand the difficulties—the many hard steps required to go from concept or laboratory scale to full-scale, economically feasible operation. Now is the time for engineers to speak up, to speak truth to power.

Ethical issues related to nuclear power abound concerning proliferation; the pros and cons of reprocessing; questions about breeder reactors; and questions about the development of new nuclear weapons. The September 2003 issue of *Nuclear News* brought out some of the issues involving the ethics of engineering related to nuclear power. "[S]ome activists have demanded that transmission lines be buried underground. They seem to be unaware that subsurface installation . . . actually brings the lines closer to nearby residents and workers" (Taylor, 2003). John Deutch, MIT University Professor, said that "Taking nuclear power off the table as a viable alternative will prevent the global community from achieving long-term gains in the control of carbon dioxide emission" (Nuclear News, 2003a). Regarding Indian Point, a nuclear power station on the Hudson River north of New York City, "Opponents of Indian Point, which include some state and local lawmakers and activist organizations, want the plant closed down for safety reasons." FEMA (Federal Emergency Management Agency) concluded, "After carefully considering all available information, we have reasonable assurance that appropriate protective measures to protect the public health and safety of surrounding communities can be taken and are capable of being implemented in the event of a radiological incident at the Indian Point facility." However, "Indian Point's home county, Westchester County, refused to submit documentation to the state pertaining to the emergency plan..." (Nuclear News, 2003b).

A report by the Progressive Policy Institute on the Bush administration's performance on homeland security also relates to nuclear power plants. "The Nuclear Regulatory Commission, in reaction to the Sept. 11 attacks, quickly issued heightened security regulations for all nuclear power plants.... If anything, the NRC could be faulted for overkill, as nuclear power plants have always been extremely secure...." Nuclear plants received the only grade of A in the report. The overall administration grades were C or D (Nuclear News, 2003c). The French government eased restriction on thermal releases from nuclear power plants in early August to allow Electricité de France (EdF) to safely maintain the grid while satisfying the soaring demand during a heat wave affecting most of Europe. "Since nuclear power provides some 80 percent of France's electricity, however, measures had to be taken to ensure that EdF could rely on its nuclear fleet. . . . Soon after the exemptions were announced, antinuclear organizations attacked the government, claiming that the measures were taken to help the nuclear industry. The media soon shifted their attention, however, when the estimate of total heat-related fatalities had increased significantly, reaching more than 10,000 on August 21" (Nuclear News, 2003d).

In comments about possible budget cuts, a DOE official said, "The larger point is that the closure of the MIT [research] reactor, which is viewed internationally as the most important nuclear engineering program and research reactor in the world, would send a negative signal to the nuclear engineering community" (Michal, 2003).

"Editors have an important role in reporting news about nuclear energy . . .

[I]n *The Augusta* [Georgia] *Chronicle,* news articles about the Savannah River Site often had a skull-and-crossbones icon imbedded in them. The words 'radiation' and 'plutonium' were usually preceded by the adjectives 'lethal' or 'deadly'" (Reinig and McKibben, 2003).

In the United Kingdom, the Health and Safety Executive (HSE) published a document on nuclear safety regulation that said it "accepts that good safety and good commercial performance are both factors of good management. The document warns, however, that there is potential for tension between them" (Nuclear News, 2003e).

People on both sides have a tendency to exaggerate to counter an opponent who is exaggerating. But, that is wrong and self-defeating in the long run. Once credibility is lost, it is hard to regain—and may not, in fact, be able to be regained. The debate on nuclear energy and other energy issues would be enhanced if knowledgeable professionals would clarify the issues and separate fact from opinion. People in the technical community who understand these technologies have a duty to address them objectively.

Of course, engineers are citizens, and they have the same right to voice their opinions as other citizens. However, once someone identifies himself or herself as an engineer, he or she puts on a mantle of careful analysis and objectivity, which should not be misused. This is a heavy burden and, unfortunately, it is frequently not accepted.

OBSTACLES TO OBJECTIVE ANALYSIS

Here are some examples of obstacles (or barriers) to the objective analysis of technical issues related to complicated problems:

- A lack of understanding of the technology is often coupled with overconfidence. Two examples are the Three Mile Island accident and the destruction of the Chernobyl reactor.
- There may be pressures from above to suppress unpleasant information, as there was in the Challenger episode.
- Engineers do not always identify fragilities in a system, or, if they do, they do not always communicate their concerns to those in power. This may have been the case in the large blackout in the Northeast last summer.

The biggest problem in engineering ethics may be the difficulty of speaking truth to power. ". . . [S]ome government researchers will face a different ethical challenge: 'to speak truth to power'. [As Lewis Branscomb wrote,] 'The users of our results, the decision makers who need our advice, will always press us to be more sure of ourselves than our data permit, for it would make their jobs easier.' This is one challenge the government researcher faces: to insist on an accurate description of what is known and what is not, to include uncertainty in the estimates, and to be clear just how far the [technology and] science can take you. A more difficult challenge comes when the [professional's] position, based on his or her research, contradicts a strongly held position of senior political appointees. These situations, while perhaps rare, can place the government researcher in a dilemma: acquiesce or leave" (Ahearne, 1999, p. 42).

> Industry researchers also face the challenge of bringing results which differ from the company's desires and, if necessary, must be willing to disagree in public. Perhaps the most publicized example in recent years in the United States [was] the silence of researchers employed by the tobacco industry, as the industry publicly claimed that the evidence was inconclusive linking smoking to lung cancer (Ahearne, 1999, p. 43).

> Studies in the ethics of large and powerful corporations have long attempted to identify the specific structures behind ethical failures. While results of the studies differ somewhat, the list of root causes frequently revolves around three structures: (1) the loss of an outward-focused organizational purpose; (2) the failure of effective concern for diverse stakeholders; and (3) the suppression of internal dissent (Branick, 2003, p. 8–9).

> High ethics firms inevitably have some form of 'open-door policy' where no threat of punishment hovers over those who are willing to report what they perceive to be wrongdoing or just plain stupidity (Branick, 2003, p. 10).

As a last resort, when lives are in jeopardy, a professional must go public, which will most likely end the professional's career. Ethical behavior can be a hard road to follow.

CONCLUSION

Norm Augustine, retired chairman and CEO of Lockheed Martin, has written about the ethical challenges facing engineers: ". . . engineering has a great deal to do with ethics, and most of the engineers whom I have seen get into trouble on ethical matters did so not because they were not decent people but because they failed to recognize that they were confronting an ethical issue." Augustine notes that "the things engineers do have consequences, both positive and negative, some unintended, often widespread, and occasionally irreversible. In fact, the ethical content of the decisions confronting engineers is increasing as the impact of their work reaches more and more people around the world" (Augustine, 2002).

Let me end with the favorite quote of the late Dr. Edward Obert, a long-time professor of mechanical engineering at the University of Wisconsin-Madison. The quote is from Socrates: "When my sons grow up, I would ask you, my friends, to punish them if they care about anything more than virtue."

REFERENCES

Ahearne, J.F. 1999. The Responsible Researcher: Paths and Pitfalls. Research Triangle Park, N.C.: Sigma Xi, The Scientific Research Society.

Augustine, N.R. 2002. Ethics and the second law of thermodynamics. The Bridge 32(3): 4–7.

Branick, V.P. 2003. Schooled by scandals. America 189(6): 8–10.

Deutch, J., E. Moniz, S. Ansolabehere, M. Driscoll, P. Gray, J. Holdren, P. Joskow, R. Lester, and N. Todreas. 2003. The Future of Nuclear Power: An Interdisciplinary MIT Study. Cambridge, Mass.: Massachusetts Institute of Technology. Also available online at http://web.mit.edu/nuclearpower/.

Glanz, J., and J. Schwartz. 2003. Dogged Engineer's Effort to Assess Shuttle Damage. The New York Times, September 26: A1, A16.

Holly, C. 2003. EPA's New NSR Rule: Common Sense Change or Massive Loophole? Energy Daily, August 28, 2003.

Kirsen, S. 2003. Wind Power's New Current. New York Times, August 28, A1, A16.

Michal, R. 2003. Gutteridge: on the DOE's assistance to university nuclear engineering programs. Nuclear News 46(10): 22–26.

NRC (National Research Council). 1989. Improving Risk Communication. Washington, D.C.: National Academy Press.

——. 1992. Ground Water at Yucca Mountain: How High Can It Rise? Washington, D.C.: National Academy Press.

——. 1995. Technical Bases for Yucca Mountain Standards. Washington, D.C.: National Academy Press.

——. 1996. Understanding Risk: Informing Decisions in a Democratic Society. Washington, D.C.: National Academy Press.

——. 2001. Disposition of High-Level Waste and Spent Nuclear Fuel: The Continuing Societal and Technical Challenges. Washington, D.C.: National Academy Press.

Nuclear News. 2003a. MIT report: seek 1,000 reactors by 2050. Nuclear News 46(10): 12.

——. 2003b. FEMA, NRC approve site's emergency plan. Nuclear News 46(10): 15.

——. 2003c. Top grade for nuclear plants; lower for materials. Nuclear News 46(10): 27.

——. 2003d. Late news in brief. Nuclear News 46(10): 17.

——. 2003e. U.K. nuclear regulator presents new strategy. Nuclear News 46(10): 45.

PCAST (President's Committee of Advisors on Science and Technology). 1997. Report to the President on Federal Energy Research and Development for the Challenges of the Twenty-First Century. Report of the Energy Research and Development Panel. Washington, D.C.: Office of Science and Technology Policy, Executive Office of the President. Also available online at *http://www.ostp.gov/PCAST/pch0exez_all.htm.*

Reinig, W.C., and J.M. McKibben. 2003. Citizen advocacy for nuclear energy: a road less traveled. Nuclear News 46(10): 28–30.

Spohn, W.C. 2003. Building bridges. America 189(6): 23.

Taylor, G.M. 2003. Power to the people. Nuclear News 46(10): 4.

State of the Art in Engineering Ethics

Methodologies for Case Studies in Engineering Ethics

CHARLES E. (ED) HARRIS
Texas A&M University

The methodology presented in this paper has two aspects: analytical and problem-resolution. The analytical aspect suggests concepts for identifying the types of issues in a case—factual issues, conceptual issues, application issues, and moral issues. The problem-resolution aspect involves "bottom-up" techniques and "top-down" techniques. Bottom-up techniques rely on moral intuitions rather than moral theories. These methods include weighing, casuistry, and finding a creative middle way. Top-down methods appeal to a general moral theory and are sometimes useful in applied ethics. Both methods are familiar in Western philosophy as utilitarianism and the ethics of respect for persons.

Most education in ethics and professional responsibility relies heavily on case studies. This is true of medical, legal, nursing, veterinary, dental, and business ethics. It is also true of engineering ethics. Students in my large classes in engineering ethics (approximately 600 each semester) often tell me that their favorite part of the course is the case studies, reflecting the practical orientation that characterizes all professionals. The ethical and professional concerns of people who defend clients in court, treat people who are sick, manage companies, fill teeth, operate on pets, and design bridges can best be addressed by way of cases that focus on activities relevant to their usual activities.

I find it useful to divide cases into three categories: micro-cases, macro-cases, and exemplary cases. Broadly defined, micro-cases are cases in which an individual professional makes decisions involving ethical or professional concerns. These decisions may have a limited impact or a wide-ranging impact. For example, John must decide whether he will accept a rather large gift from a supplier. Alison must decide whether she is going to take part in a project that is environmentally destructive.

Macro-cases typically involve social policies, legislation, governmental administrative decisions, or the setting of policies for professional societies. In engineering, these policies usually have to do with the impact of technology on society. How should privacy be protected with respect to computers? How should computer crimes be treated? What kind of intellectual property rights should be granted to the creators of software? What policies should engineering societies adopt with respect to the environment? Should the cloning of human beings be pursued?

Exemplary cases involve situations in which professionals act in an admirable way in their professional capacities. Exemplary cases have two characteristics. First, decisions have already been made and a course of action already taken. In other words, no dilemma remains to be resolved. In exemplary cases, the dilemma has already been resolved in an exemplary way. Second, the behavior exhibited is praiseworthy, either because it is a paradigm of right action or because the action is taken in the face of adversity or because the action goes beyond what might be considered required under the circumstances. Exemplary cases can involve micro- or macro-issues.

Here is an example of a micro-case involving exemplary action. In the late 1930s, a group of General Electric engineers spent time outside their normal working hours to develop the sealed-beam headlight. Apparently, the prevailing consensus was that the headlight was not technically feasible. Nevertheless, the engineers accomplished their task. Sometimes, an engineer who simply performs what appears to be his or her professional duty can also exhibit exemplary action. Roger Boisjoly, an engineer who protested the launch of the Challenger at considerable risk to his career, exhibited exemplary action.

METHODS OF ANALYSIS

Methods of analysis can be used to identify the types of issues involved in a case: factual issues, conceptual issues, application issues, and moral issues.

Factual Issues

A factual issue has two characteristics: (1) it is a disagreement over a matter of fact, and (2) this matter of fact is crucial to resolving the problem. A fact, unlike a factual issue, is a matter that has already been settled and is uncontroversial. Factual issues arise, for example, in cases in which we do not know how much a certain modification in a design will cost or what the effects of a certain course of action will be or how accurate a given test is or what risks are involved in a certain technology.

In the real world, empirical research should be used to resolve a factual issue. Some factual issues, however, cannot be resolved by investigation. Some technological questions cannot be answered, such as questions about consequences that can only be answered in the future. In these cases, the most realistic approach is to leave the factual question unanswered and make a decision in the context of factual uncertainty. Especially in the classroom, it is not appropriate to make assumptions that resolve an issue in a way that could not be done in a real-world context.

Here is a case involving a factual issue. A new law requires that the lead content of drinking water be less than 1.0 part per billion (ppb). Melissa is a safety engineer who has tested her company's drinking water by two methods. Method A gives a reading of 0.85 ppb; Method B gives a reading of 1.23 ppb. She must fill out a government report describing the quality of her company's water. If the lead content exceeds 1.0 ppb, her company will be fined. She must decide whether to report the results of Method A or Method B. In this case, her decision is based primarily on the factual issue of which method is the most accurate.

It is important to keep in mind that many controversies that appear to be about moral issues are traceable primarily to disagreements over facts. Two people may disagree about the proper course of action because they disagree about the consequences of a given course of action. Two engineers may disagree about which of two designs is ethically more acceptable because they disagree about which one is safer. They may agree on the moral parameters of the case, namely that the safest design should be chosen, but they may disagree over which design is safer. Although such a disagreement might be called a moral or ethical disagreement, it is really a disagreement over factual issues, unless they disagree over the definition of "safe." Engineering students are often inclined to say that ethics are "soft" (in cases where a factual disagreement cannot be settled). It is important, therefore, to realize that sometimes, even though moral parameters may be agreed upon, there may be irresolvable disagreements over facts.

Conceptual Issues

A conceptual issue is (1) a disagreement over a definition of a concept that is (2) crucial to resolving a problem. Two engineers may differ over whether a design is safe because they have different definitions of (i.e., criteria for) "safe." They may disagree about whether a given action is a conflict of interest because they may have different definitions of "conflict of interest." They may disagree over whether something is a bribe because they have different conceptions of a bribe and how to distinguish one from extortion or "grease money."

Here is an example of a case involving a conceptual issue. Sally is a mechanical engineer employed by General Motors to design automotive gas tanks. According to government safety standards, the automobile must be able to survive a "moderate impact" with no chance of the gas tank catching fire. In recent tests, in cars that crashed at 35 miles per hour (mph) the gas tanks did not catch fire, whereas in 20 percent of cars that crashed at 45 mph they did. She knows she must first determine how the government defines "moderate impact."

Probably the most effective way to come up with a definition is to derive one from paradigm, or standard, cases. A paradigm case of a bribe might be one in which an engineer accepts a large sum of money to specify a product that is not the most appropriate one for the design. From this standard case, we might derive a working definition of a bribe as an offer of something of value to induce a person to perform an action that is morally inappropriate to his or her office or role. If definitions differ, it may be possible to argue that one definition is more in accord with standard practice or paradigms or that one definition is more useful or easier to apply. If there are continuing differences over conceptual issues, the important thing is to be aware of the differences.

Another important consideration is whether a concept is "moralized" or "nonmoralized." A moralized concept includes an implicit moral judgment that the action to which the concept refers is either morally acceptable or unacceptable. When we label something as a bribe, we make a presumptive judgment that it is wrong, because, as we have seen, we usually define bribery as giving something of value to induce a person to perform an action that is *morally inappropriate* to his or her office or role. Breaking confidentiality, for example, is *prima facie* morally wrong, because we define it as violating a commitment or breaking a rule that is morally justified.

Of course, the fact that an action is a bribe makes only a presumptive case that it is morally wrong. There might be a moral consideration that overrides the fact that we are giving a bribe. Bribing a Nazi guard to get your grandmother out of a concentration camp would be morally permissible, because the office of a concentration-camp guard is itself morally illegitimate. Breaking confidentiality is *prima facie* bad, but it may be justified when the safety of the public is at stake.

Some concepts, by contrast, appear to be morally neutral. We may call them nonmoralized concepts. In deciding whether computer software is a work of authorship (like a book) or an invention (like a machine), we must define "work of authorship" and "invention." These definitions do not appear to involve moral judgments about the value of these two types of creative products.

Application Issues

An application issue is a question of whether or not a concept applies to a given situation. An application issue is (1) a disagreement over the application of a concept in a particular situation that is (2) crucial to resolving a problem. I just referred to the question of whether computer software should be classified as a work of authorship or an invention. This is an application issue, because the question is whether the concept of a work of authorship (once we have defined it) or the concept of invention (once we have defined it) best applies to software. Of course, neither of these concepts applies particularly well, and this is characteristic of application issues. An application *issue* is one in which we have trouble deciding whether a concept applies in a given situation. We have no trouble deciding whether killing a person by stabbing him in the back to get his money is murder, but we do argue over whether euthanasia is murder. Similarly, engineers might argue over whether attending a conference in Hawaii sponsored by a vendor is a bribe, or whether giving one client general information about another client's projects is a breach of confidentiality.

Here is an example of an application issue. Larry is an aerospace engineer who is a member of the Quaker religion, which is committed to nonviolence. Larry was hired by his firm to design passenger airplanes, but his boss has recently reassigned him to design military fighters. Larry must decide whether to accept the new assignment or quit and find a new job. He must decide whether his commitment to "nonviolence" requires not only that he refrain from operating military aircraft, but also that he refrain from designing them.

Application issues often arise in the law. The Constitution requires that citizens be given a "speedy" trial. If a citizen is kept in jail for two years without a trial, is this a denial of his constitutional right to a speedy trial? A city has a law against "vehicles" in the park, and a child rides a skateboard into the park. Is a skateboard a "vehicle"?

Moral Issues

A fourth type of issue is a genuine moral issue, usually a conflict between two or more values or obligations. Engineer Tom does not want to give the customs officer money, but he needs to get something through customs to complete a project that is important for the local economy as well as for his firm. Here Tom faces a conflict between his obligation not to pay bribes or grease money and his obligation to complete the project. Engineer Jane is not sure whether she should design a slightly safer product that will be considerably more expensive for consumers. Jane faces a conflict between her obligation to produce safe products and her obligation to produce inexpensive products.

Here is another example of a moral issue. Harry works for a large manufacturer in the town of Lake Pleasant. His company employs half of the people in the town, which is in an otherwise economically depressed part of the country. Harry discovers that his company is dumping chemicals into the local lake that may pose a health hazard. The lake is the town's main source of drinking water. Harry is told that the company dumps these chemicals into the lake because disposing of them in any other way would be so expensive that the plant would have to close. Should Harry report his company's practice to the local authorities? Harry faces a conflict between his obligation to the health of the citizens of Lake Pleasant and his obligation to the economic welfare of the citizens of Lake Pleasant.

BOTTOM-UP METHODS OF PROBLEM RESOLUTION

Sometimes moral conflicts remain even after all of the factual, conceptual, and application issues have been resolved. Therefore, we should consider some methods for resolving moral conflicts. Following a nomenclature often used in medical ethics, I find it useful to divide methods of resolving conflicts into bottom-up and top-down methods. Bottom-up methods start on a fairly concrete level, close to the details of the case, and work toward a solution. These methods adopt generally-accepted, intuitively plausible moral concepts that are a part of the moral thinking of most people, at least in our society. They work on what R.M. Hare, a prominent moral philosopher, would call the intuitive level of moral thinking (Hare, 1981).

Weighing or Balancing

The simplest bottom-up method might be called balancing or weighing. Reasons for alternative evaluations are considered, or "weighed," and the alternative with the most convincing reasons is selected. We examine the reasons for and against universal engineering registration and, all things considered, find one set of reasons more convincing than the others. If we find the reasons on both sides equally convincing, either option is morally permissible.

Engineer Jane, who owns a civil engineering design firm, has a chance to bid on part of the design work for a fertilizer plant in Country X. The plant will increase food production in a country where many people do not have sufficient food. Unfortunately, the plant will have some bad environmental effects, and correcting the problems will make the fertilizer more expensive, too expensive for farmers in Country X. Should she bid on the design? She may decide to list considerations in favor of submitting a bid and considerations against it. On the one hand, she will be contributing to the saving of many lives, the economic development of Country X, and the economic advancement of her firm. On the other hand, she will be contributing to the environmental degradation of Country X, and her firm may receive some negative publicity. She must attempt to balance these two sets of considerations and determine which has the greater moral "weight." Balancing does not provide specific directions for comparing alternative courses of action, but sometimes such direction is not necessary.

The Method of Casuistry or Line Drawing

The second method is casuistry, or what I call line drawing. Although the method I have developed for students is more formal than would ordinarily be used in real-world situations, I believe the underlying ideas are what we might call moral common sense. Casuistry has a long history in the moral tradition of the West, going back at least to Cicero. Recently, casuistry has been used to make decisions in medical ethics. Congress established the National Commission for the Protection of Human Subjects of Biomedical Research in 1974. Deep religious and philosophical differences between members of the commission made progress difficult until the group decided to talk about specific examples of morally objectionable experiments ("paradigm cases"). The members found that they could agree on the characteristics ("features") of these experiments that made the experiments wrong. Some members of the commission recognized that they were using the ancient technique of casuistry, and the method subsequently came to be accepted in medical ethics cases.

In casuistry, a decision about what to do or believe in a problematic situation is made by comparing the problematic situation with a clear situation. The comparison—reasoning by analogy—is made by comparing the features of the test case with the features of a "positive paradigm case" and a "negative paradigm case." A feature is a characteristic that distinguishes a paradigm case from the test case, the subject of the analysis. A negative paradigm is a clear or uncontroversial example of an action that is wrong or morally impermissible; a positive paradigm is a clear and uncontroversial example of an action that is right or morally permissible.

Casuistry, or line drawing, can be used to resolve two distinct kinds of questions. First, it can be used to resolve an application issue, for example, to determine whether an action really constitutes a bribe. Second, it can be used to resolve a moral issue, for example, once we have determined that an offer really is (or is not) a bribe, whether or not we should accept it or offer it. Of course, in most circumstances, a bribe should not be accepted or offered, but offering or accepting a bribe might be justifiable in a few cases. To cite an earlier example, during World War II, if I could have bribed a Nazi guard to get my grandmother out of a concentration camp, I might decide that offering a bribe is justifiable.

The following example illustrates how casuistry can be used to settle an application issue and to settle a moral issue. Denise is an engineer at a large construction firm. Her job requires that she specify rivets for the construction of a large apartment building. She has the power to make the decision by herself. After some research and testing, she decides to use ACME rivets for the job, because, indeed, they are the best product. The day after she orders the rivets, an ACME representative visits her and gives her a voucher for an all-expense paid trip to the ACME Technical Forum in Jamaica. The voucher is worth $5,000, and the four-day trip will include 18 hours of classroom instruction, time in the evening for sightseeing, and a day-long tour of the coastline. The time will be roughly divided between education and pleasure. Does this trip constitute a bribe? A line-drawing analysis might look like Table 1.

In a line-drawing analysis, one must decide not only where to place the "x's" on the spectrum, but also how much "weight" or importance to give each "x." Some features may be more important than others. For example, one might decide that because the offer was made *after* the decision to buy ACME rivets the

TABLE 1 Line-Drawing Analysis for Resolving an Application Issue

Features	Positive Paradigm	Test Case	Negative Paradigm
Gift Size	$1.00	_ _ _ _ _ _ _X_ _	$ 5,000
Timing	After decision	X_ _ _ _ _ _ _ _ _	Before decision
Reason	Education	_ _ _ _X _ _ _ _	Pleasure
Power to make decisions	With others	_ _ _ _ _ _X _ _	Alone
Quality of product	Best	_X_ _ _ _ _ _ _	Worst

gift cannot be considered a bribe. It may be a bribe, however, to *other* engineers, who may believe that buying ACME products results in offers of nice trips. However, to Denise it is certainly not a paradigm bribe.

Line-drawing analysis can also be used to determine whether Denise should take the trip. Even if she decides the trip is not a bribe, she might still decide not to accept the offer. The features important to this decision may be different from the ones in the first analysis, although there may be some overlap. In the second analysis, it will be important to consider the influence of the gift on future decisions by Denise and other engineers, the company policy on accepting gifts, and the appearance of bribery if the gift is accepted. Some features from the first analysis, such as the educational value of the technical forum, would be relevant here too. Table 2 is a line-drawing analysis to resolve the moral question of whether Denise should accept the offer.

According to the analysis in Table 2, the issue is not clear. However, the problems associated with accepting the gift are serious enough that Denise probably should not accept it. In the next section, I shall suggest conditions under which accepting the gift would probably be morally permissible.

TABLE 2 Line-Drawing Analysis for Resolving a Moral Issue

Features	Positive Paradigm	Test Case	Negative Paradigm
Influence on future decisions	None	_ _ _ _ _X _ _	Great
Company policy	May accept	_ X _ _ _ _ _ _	May not accept
Appearance	No problem	_ _ _ _ _X _ _	Appearance of a bribe
Educational value	Great	_ _ X _ _ _ _ _	Minimal

But first, here are some concluding thoughts about the method of casuistry. In general, the more features that are included in an analysis, the better. For the sake of simplicity, I used only four or five, but the more features you include, the more helpful and accurate the analysis becomes.

Casuistry is an inherently conservative method. In arriving at paradigm cases for comparison with test cases, we assume that our intuitive, common sense moral judgments are correct. This assumption is usually valid, but not always, particularly in areas where morality is changing or when the case involves a novel experience. It might be difficult to find uncontroversial paradigm cases for some issues in environmental ethics, for example.

For casuistry to work well in the context of a profession, the professional community must agree on paradigms of acceptable and unacceptable practice. Engineers must agree on paradigmatic examples of acceptable and unacceptable practice with respect to conflicts of interest, confidentiality, and other issues. In the area of medical ethics, for example, there is now widespread agreement about whether actions taken in certain publicized cases were moral or not. These agreed-upon bench marks can then be compared to more controversial cases. I believe there has been less discussion of bench mark cases in engineering.

Creative Middle Ways

A third method of resolving a problem is finding a creative middle way. Suppose there is a conflict between two or more legitimate moral obligations and that two of them appear to be at loggerheads. Sometimes by creative thinking, it is possible to find a course of action that satisfies both, although perhaps not in the way that was originally supposed. For example, a plant might be emitting some dangerous pollutants that are environmentally harmful, but completely eliminating them would be so expensive that the plant would have to close, throwing many local inhabitants out of work. Assuming there is an obligation both to preserve jobs and to protect the environment, a creative middle way might be to eliminate the worst pollutants and forego a complete cleanup until more economical means of doing so can be found. This alternative would be particularly attractive if the remaining pollutants would not cause irreversible damage to the environment or to human health.

This solution, and most creative middle-way solutions, involves compromise. Environmentalists might not be completely satisfied with this solution because not all of the pollutants will be removed. Plant managers might not be completely satisfied because the solution will still involve considerable expenditures for pollution control. Nevertheless, environmentalists will accomplish something, and the plant owners can remain in the town and even build up a considerable amount of public good will.

In the line-drawing analysis presented in the previous section, there might also be a creative middle way. Suppose we take the two competing values: (1) the educational and recreational value of the trip; and (2) avoiding the appearance of bribery and undue influence on professional judgment. Denise's manager might suggest: (1) that she take the trip but that the company pay her expenses; and (2) that engineers who were not involved in the decision also be allowed to take the trip. Furthermore, it must be understood that company engineers will be allowed to attend the forum, at the company's expense, whether or not the company buys ACME products. This arrangement would only make sense, of course, if the forum is of very great technical value. This solution would allow Denise to honor competing obligations in a creative way.

Two limitations of this method come to mind. First, sometimes there is no creative middle way, even if it is desirable. In the example cited above, all of the pollutants may be so damaging to the environment that no half-way measures will work. Furthermore, there might not be a way to do the cleanup more economically. In that case, the plant might just have to close. In the line-drawing example, Denise's company might not be able to pay her expenses. A second limitation is that sometimes the creative middle way is not morally appropriate. Sometimes one of the options is so morally repugnant that we must choose the other one. Still, a creative middle way is often a good solution to a complex, practical moral problem.

TOP-DOWN METHODS OF RESOLUTION

In some cases, the appeal to moral common sense may not be sufficient. In those cases, it may be useful to appeal to more fundamental moral ideas, such as those developed in philosophical theories. Although the role of moral theory in applied or practical ethics is controversial, I believe moral theorists have attempted to find fundamental moral ideals that can generate or explain all or most of our common-sense moral ideas. This goal has been only partially achieved, because there are at least two prevalent moral theories today, and neither one can explain the fundamental ideas of common morality in a completely satisfactory way. These two theories are utilitarianism, usually associated with Jeremy Bentham and John Stuart Mill, and the ethics of respect for persons, usually associated with Immanuel Kant. The main idea behind utilitarianism is to maximize overall human well-being; and the main idea behind the ethics of respect for persons is to respect the rights and moral agency of individuals.

Although the existence of two theories rather than one may be an embarrassment to theorists, practical ethicists can take a more positive attitude because the conflict between the ideas behind these two theories often arise in real-world moral controversies. Common morality, at least in the West, may not be a seamless web. In fact, it may be composed of two strands: (1) considerations having to do with utility, or the well-being of the greatest number of people; and (2) considerations having to do with justice and the rights of individuals.

An understanding of moral theory could serve several functions in practical ethics. First, the two perspectives can often be helpful for identifying and sorting out different types of arguments and for recognizing that different types of arguments have deep moral roots. In arguments for and against strict protections for intellectual property, for example, knowing that some arguments are utilitarian can be helpful. From the utilitarian perspective, protecting intellectual property promotes the flourishing of technology and, thereby, the good of society. Utilitarian arguments can also be made that strong protections for intellectual property limit the sharing of new ideas in technology and are thereby detrimental to the general good. Arguments from the respect-for-persons perspective often focus on the individual's right to control, and reap the profits from, the fruits of his or her own labor, regardless of the impact on the larger society.

Second, understanding these fundamental, yet divergent, moral perspectives often enables an ethicist to anticipate a moral argument. Just thinking about the two theories and the kinds of arguments they would support could have led one to expect that some arguments regarding intellectual property would take the utilitarian approach and others would take the rights-of-ownership approach.

Third, familiarity with these two perspectives can sometimes help in determining whether there has been closure on a moral issue. If arguments from both perspectives lead to the same conclusions, we can be pretty confident that we have arrived at the right answer. If the arguments lead to different conclusions, the discussion is likely to continue. When different conclusions are reached, there is no algorithm, unfortunately, for deciding which moral perspective should prevail. In general, however, the Western emphasis on individual rights and respect for persons takes priority, unless harm to individuals is slight and the utility to society is very great. With these considerations in mind, we can now look at the two moral theories.

The Ethics of Utilitarianism

A principle of utilitarianism is that the right action will have the best consequences, and the best consequences are those that lead to the greatest happiness or well-being of everyone affected by the action. Consider the following case. Kevin is the engineering manager for the county road commission. He must decide what to do about Forest Drive, a local, narrow, two-lane road. Every year for the past seven years, at least one person has crashed a car into trees close to the road and been killed. Many other accidents have also occurred, causing serious injuries, wrecked cars, and damaged trees. Kevin is considering widening the road, which would require that 30 trees be cut down. Kevin is already receiving protests from local citizens who want to protect the beauty and ecological integrity of the area. Should Kevin widen the road?

In this case, the conflicting values are public health and safety on the one hand and the beauty and ecological integrity of the area on the other. Let us suppose that widening the road will save one life and prevent two serious injuries and five minor injuries a year. Not widening the road will preserve the beauty and ecological integrity of the area. Even though the preservation will increase the happiness of many people, the deaths and injuries are far more serious negative consequences for those who experience them. Therefore, the greatest total utility is probably served by widening the road.

Cost/benefit analysis is a form of utilitarianism. I sometimes refer to it as "utilitarianism with the numbers." Instead of maximizing happiness, the focus is on balancing costs and benefits, both measured in money, and selecting the option that leads to the greatest net benefit, also measured in money. Consider an earlier case. ACME manufacturing has a plant in the small town of Springfield that employs about 10 percent of the community. As a consequence of some of its manufacturing procedures, the ACME plant releases bad-smelling fumes that annoy its neighbors, damage the local tourism trade, and have been linked to an increase in asthma in the area. The town of Springfield is considering issuing an ultimatum to ACME to clean up the plant or pay a million-dollar fine. ACME has responded that it will close the plant rather than pay the fine. What should Springfield do?

A cost/benefit analysis might show the costs of and benefits of not levying the fine and keeping the plant open (Table 3) and or levying the fine and losing the plant (Table 4).

According to these analyses, the economic consequences of fining ACME would be much greater than the consequences of not fining ACME. Thus, the fine should not be levied.

There are two major problems with utilitarianism. One is that an accurate analysis requires a lot of factual information. This is especially evident in the cost/benefit analyses above. One must know the amounts to assign to the various costs and benefits. Even in an analysis that is not done in the cost/benefit way, the consequences of various courses of action must be known before the course of action that will have the greatest overall utility can be known. A second problem

TABLE 3 Cost/Benefit Analysis of Not Levying the Fine

Costs:	
Health expenses	$1,000,000
Nuisance odor	$50,000
Decline in housing values	$1,000,000
Decline in tourism	$50,000
Benefits:	
Wages	$10,000,000
Taxes	$2,000,000
Total	+$9,900,000

TABLE 4 Cost/Benefit Analysis of Levying the Fine

Costs:	
Loss of wages	$10,000,000
Loss of tax revenue	$2,000,000
Decline in housing values	$2,000,000
Benefits:	
Fine	$1,000,000
Increase in tourism	$50,000
Health savings	$900,000
Total	–$12,050,000

is that a utilitarian analysis can sometimes justify unjust consequences. For example, a decision not to force the plant to stop polluting will result in some people getting sick, even though overall utility will be maximized. These problems suggest that a complete analysis should include the ethics-of-respect principle.

The Ethics of Respect for Persons

From the utilitarian point of view, harm to one person can be justified by a bigger benefit to someone else. In the ethics of respect for persons, there are some things you may not do to a person, even for the benefit of others. The fundamental idea in the ethics of respect for persons is that you must respect each person as a free and equal moral agent—that is, as a person who has goals and values and a right to pursue those values as long as he or she does not violate the similar rights of others.

As this formulation suggests, the ethics of respect for persons emphasizes the rights of individuals, which are formulated, among other places, in various United Nations documents. Individual rights include the right to life and to the security of one's person, the right not to be held in slavery, the right to freedom of thought and expression, and so forth. The problem with this formulation is that it does not give any clear indication of which rights are most important. When rights conflict, it is important to know which ones are most important.

Alan Gewirth, a contemporary philosopher, has suggested that there are three levels of rights (Gewirth, 1978). Level I, the most important rights, includes the right to life, the right to bodily integrity, and the right to mental integrity. I would add to those the right to free and informed consent to actions that affect one. Level II includes the right not to be deceived, cheated, robbed, defamed, or lied to. It also includes the right to free speech. Level III includes the right to acquire property and the right to be free of discrimination. For Gewirth, Level I rights are the fundamental rights necessary for effective moral agency. Level II rights are necessary to preserving one's moral agency. Level III rights are necessary to increasing one's level of effective moral agency. Whether or not one accepts this arrangement, most of us would probably recognize that some rights are more important than others.

Consider the following case. Karen, who has been working as a design engineer under Andy, has learned that he is about to be offered a job as head safety inspector for all of the oil rigs the company owns in the region. Karen worries that Andy's drinking may affect his ability to perform his new job and thereby endanger workers on the oil rigs. She asks Andy to turn down the new assignment, but he refuses. Should Karen take her concerns to management? In this case, Andy's right to advance his career (by trying to acquire property), which is a Level III right, conflicts with the workers' rights to life and bodily integrity, which are Level I rights. In this conflict, the rights of the workers are more important, and Karen should take her concern to management.

In arbitrating conflicts between rights, two additional issues should be kept in mind. First, there is a distinction between violating and infringing a right. A right is violated if it is denied entirely. I violate your right to life if I kill you. A right is infringed if it is limited or diminished in some way. A plant infringes on my right to life if it emits a pollutant that increases my risk of dying of cancer. Second, rights can be forfeited by violating or perhaps infringing on the rights of others. I may forfeit my right to life if I kill someone else. I may forfeit some right (perhaps the right to free movement) if I steal from others and thus infringe on or violate their right not to be robbed.

Finally, the Golden Rule is also a principle associated with the ethics of respect for persons. Most cultures have a version of the Golden Rule. The Christian version requires that we treat others as we would have them treat us. In the Islamic version, no man is a true believer unless he desires for his brother that which he desires for himself. If we consider ourselves to be moral agents, the Golden Rule requires that we treat others as moral agents as well.

There are two primary problems with the ethics of respect for persons. First, the rights test and the Golden Rule are sometimes difficult to apply. We must determine when there is a conflict of rights, which rights are most important, and whether rights have been violated or merely infringed upon. With the Golden Rule, we must assume that others have the same values we do. If they do not, treating them as we would wish to be treated may be unfair. Second, it may be justifiable at times to allow considerations of utility to override considerations of the ethics of respect for persons, especially if the infringements of rights are relatively minor and the benefit to the general welfare is great.

CONCLUSION

I have presented a number of tools for analyzing and resolving ethical problems. The important thing to keep in mind, however, is that these tools cannot be used in a mechanical way. They are not algorithms. One must decide if the issue to be resolved is really factual or conceptual, for example. One must also decide when the line-drawing method or finding a creative middle way is most appropriate and when an issue can best be approached as a conflict between general human welfare (utility) and the rights of individuals (the ethics of respect for persons). When there is such a conflict, there is no mechanical way to determine which perspective should be considered most important. In the West, we accord great importance to individual rights, but they do not always take precedence. The techniques and methods I have described are helpful for thinking about ethical issues, but they are no substitute for moral insight and moral wisdom.

REFERENCES

Gewirth, A. 1978. Reason and Morality. Chicago: University of Chicago Press.
Hare, R.M. 1981. Moral Thinking: Its Levels, Method, and Point. Oxford, U.K.: Oxford University Press.

Responsibility and Creativity in Engineering

CAROLINE WHITBECK
Online Ethics Center for Engineering and Science
Case Western Reserve University

Engineering ethics, like medical ethics, has become a branch of the larger field of practical and professional ethics. Engineers first formulated ethical norms specifically for engineering practice in the first half of the twentieth century, when many professional engineering societies developed codes of ethics for their members. Since the National Project on Philosophical Ethics and Engineering in 1978–1981, philosophers and other scholars in the humanities have also weighed in on the subject. This paper examines the notions of responsibility, which is central to engineering ethics and to professional ethics generally, and creativity, which is necessary for the exercise of responsibility. The topics addressed in this paper include: professions and professional ethics; the role of engineering experience in the development of ethical guidelines for engineers; the notion of responsibility per se; the role of synthetic or creative reasoning in the fulfillment of professional responsibility; limitations on the foresight necessary for the exercise of responsibility; and bringing engineering knowledge to bear on societal choices about technology. Some of these topics have been discussed in more detail elsewhere (Whitbeck, 1998).

PROFESSIONS AND PROFESSIONAL ETHICS

My first topic is a description of professional ethics and a discussion of how the moral requirements for engineers (and other professionals) differ from the requirements for everyone else. Two characteristics distinguish the practice of professions from other occupations: (1) the mastery of a specialized body of knowledge; and (2) the application of that knowledge to securing or preserving the well-being of others.

Professional societies play a major role in the development of ethical norms (see Herkert's contribution to this workshop, pp. 107–114 in this volume). In fact, their role in developing ethical norms is what distinguishes professional societies from disciplinary, technical, scholarly, and "learned" societies. Whereas professional societies focus on professional practice, these other kinds of societies focus on technical or scholarly advances in a specific discipline or field. The National Society of Professional Engineers is a purely professional engineering society; other engineering societies, such as IEEE, are both disciplinary and professional. Although professional engineering societies began to develop codes of ethics in the early decades of the twentieth century, and the American Chemical Society did so in the 1930s, the American Physical Society did not issue its first code of ethics until 1992 (that code deals exclusively with research ethics). Before 1992, physicists seem to have considered themselves practitioners of a discipline rather than a profession. However, there is a growing realization that research (especially publicly funded research) involves the welfare of many others, including, but not only, the subjects or participants in the research.

The American Philosophical Association is a disciplinary or learned society, and philosophers have no code of ethics, reflecting a judgment that philosophy is a discipline and not a profession. Some philosophers seem confident that they affect nothing. Philosophers who are also teachers are members of the teaching profession, however, and university professors do have a code of ethics.

THE ROLE OF ENGINEERING EXPERIENCE IN THE DEVELOPMENT OF ETHICAL CODES FOR ENGINEERS

Next, we shall consider the role of engineering experience in the development of ethical guidelines for engineers. Ethical codes and guidelines for engineers come from many sources. Philosophers have had a hand in some of them. However, the most interesting and most valuable codes are based on engineers' experiences, the problems and pitfalls they actually encounter in their professional practice.

These codes and guidelines embody the profession's accumulated wisdom about its practice, the morally significant problems that arise, and appropriate limits, priorities, and prudent measures for avoiding potential moral pitfalls. They stand closest to the Aristotelian tradition of philosophical ethics, as contrasted with top-down Enlightenment theories of ethics that attempt to deduce ethical norms from a few general principles. Engineering ethics has paid much closer attention to practical experience than at least one influential wing of biomedical ethics, which early on attempted to formulate a few abstract principles and then fit all problems and issues to those principles.

Examples of experienced-based rules that set prudent boundaries are rules that limit the value of gifts an engineer can accept from business associates and the warning against working under a commission because it might create a conflict of interest. Examples of guidance about setting priorities for responsibilities and obligations are rules that give public health and safety priority over other important values, such as maintaining client confidentiality. (Research integrity is of central concern in engineering research.)

Ethical codes and guidelines are generally "living" documents in the sense that they are revised as engineers' understanding of a moral situation evolves or as conditions of practice and the moral situations themselves change with social or technological changes. However, as I will discuss later, the rapid rate of technological and social change has made it difficult for revisions to keep pace with new problems.

THE NOTION OF RESPONSIBILITY

The notion of "responsibility" is the central moral concept in engineering ethics, and in professional ethics generally. Responsibility in the moral or ethical sense is based on the ends to be achieved rather than the acts to be performed. Responsibility typically requires the application of the specialized knowledge that characterizes a profession.

As we have seen, professions are distinguished from other occupations because the practice of a profession draws on a body of expert knowledge and is directed toward securing major aspects of well-being for others. Many ethical notions are applicable to professional ethics, but the notion that best captures the special moral situation of the practitioner of a profession is professional responsibility.

Professional responsibilities—exemplified by statements such as "engineers are responsible for the public safety" or "research investigators are responsible for the integrity of research"—require that relevant *expert knowledge* be synthesized to achieve an *end*. They require judgments that only those who have mastered such knowledge can make. Whatever moral lessons we may have learned in kindergarten, we did not and could not have learned as children how to fulfill professional responsibilities. Only adults with higher cognitive skills can learn to fulfill professional responsibilities.

The exercise of professional responsibility requires both competence and concern. The ethical dimension of practicing competently is highlighted for engineers by the rule in many engineering codes of ethics that forbids engineers to accept assignments beyond their competence.[1] Engineers working beyond their

---

[1]At the time of this writing, research investigators do not generally recognize a professional obligation to work only within the limits of their competence, and many represent the distinction between incompetent research and unethical research as an absolute distinction. However, in some areas of research, certain sorts of incompetence, such as incompetence that creates major safety

competence are, for that reason alone, considered to be acting unethically. Responsible engineering practice requires both competence and the exercise of sufficient care to bring that competence to bear on a given problem.

In contrast to specifications of ethical responsibilities, ethical rules and obligations (as well as legal and organizational rules and obligations) typically specify *acts that are forbidden or required*—for example, "do not offer or accept bribes" or "you are obligated to disclose any conflicts of interest to all parties to an agreement." Following ethical rules and meeting ethical obligations may not require the application of professional knowledge (either "knowing how" or "knowing that") other than perhaps the ability to recognize when the conditions mentioned in the rules apply—for example, what forms a bribe might take in a specific professional context.

Rules that are specific to a profession derive their moral authority from their contribution to the fulfillment of the characteristic responsibilities of the profession. For example, the stricture against abandoning a patient is specific to medicine; an engineer might sometimes be wrong to leave a project without first being assured of the presence of another engineer, but there is no general stricture against an engineer leaving a project without finding a replacement. This is because the physician-patient relationship is itself an instrument of healing, and, therefore, rupturing that relationship without finding a replacement may damage the aspect of a client's welfare that is entrusted to physicians. In contrast, the people whose safety and health are the overriding responsibility of engineers and members of the public, are generally people the engineer will never meet. Thus, the interpersonal relationship between the engineer and the people whose needs he or she must consider is not an ethical consideration.

The notion of responsibility we have been considering, the notion exemplified in "engineers are responsible for the public safety" or, as Michael Loui (1998) has argued, "[e]ngineers have a responsibility for the quality of their products," is responsibility in the ethical or moral sense. As Kathryn Pyne Addelson first observed, moral or ethical responsibility is "prospective" or "forward-looking" in contrast to "blame" and other notions that are backward-looking in that they are concerned with situations that have already occurred (Kathryn Pyne Addelson, Mary Huggins Gamble Professor Emerita of Philosophy, Smith College, personal communication). A moral responsibility specifies the ends *to be achieved*.

Responsibility is sometimes used in the causal sense, as in "the storm was

---

hazards to the public or people working in the laboratory, have long been recognized as derelictions of responsibility by organizations such as the American Chemical Society. I am gratified to see that the federal definition of research misconduct now recognizes that reckless, as well as intentional, behavior may be considered research misconduct.

responsible for (i.e., caused) three deaths and millions of dollars in property damage." In this sense, responsibility may not have ethical significance—for example, when the causal agent is something, such as a storm, that is not a moral agent. If a causal agent is also a moral agent, that is, one who is *capable* of acting morally, then the causal agent usually bears some moral responsibility for dealing with the situation s/he has created. Considered by itself, however, responsibility in the causal sense is not an ethical notion.

Responsibility is also used in a third sense as a synonym for being *accountable*; in this sense, responsibility simply specifies to whom a rational agent must answer. For example, "the CEO is responsible (i.e., accountable) to the board." (Notice that when responsibility is used in this sense the phrase is always "responsibility *to*.") Responsibility in the sense of accountability applies only to rational agents. It does not specify what is required of the agent, but only *who* will judge the adequacy of the agent's actions.

In addition, there is what John Ladd (1970) has called "official responsibility," which is limited to what one is charged to do as a result of holding a particular job or office within an organization. As Ladd has argued, *official* or organizational responsibilities differ significantly from professional responsibilities and other *moral* responsibilities in that official responsibilities attach to job categories and impersonal roles rather than to particular people in particular circumstances with particular histories and human relationships who are subject to the moral demands that they carry with them. Furthermore, official or organizational responsibilities, unlike moral responsibilities, are "alienable," that is, an official responsibility is *fully* transferable from one person to another so that the first no longer has it. Notice that the official responsibilities in a job description could conceivably require unethical behavior. Nevertheless, the person holding the job still has a professional responsibility to draw attention to safety problems.

Professions claim to be autonomous, that is, that only members of the profession can establish and administer the standards that govern the practice of their profession, because people outside the profession cannot judge the quality of professional performance. For example, I may know that a surgeon ought not to leave surgical instruments inside patients, but even if I were able to monitor a surgeon's actions, even guide the surgeon's hand, it would not substitute for a trustworthy surgeon, because I would not know what to do. Because people outside the profession cannot judge professional performance, external regulation, although sometimes necessary, is a poor substitute for having trustworthy (i.e., responsible) professions and professionals.

RESPONSIBILITY AND CREATIVE/SYNTHETIC REASONING

The fourth topic is the role of synthetic or creative reasoning in the fulfillment of professional responsibility. In the exercise of professional responsibility, creative reasoning must be used to bring expert knowledge to bear on specific problems. In several respects, solving ethical problems, which is required for the exercise of professional responsibility, is analogous to solving problems of engineering design. Just as a designer needs creative abilities that a critic of design does not need, being a responsible professional requires more than judging ethical behavior.

Obeying moral rules, fulfilling obligations, and respecting others' rights typically require little professional knowledge because they provide explicit descriptions of the required actions. For example, the rule that "engineers have an obligation not to disclose a client or employer's proprietary information" specifies what engineers must avoid doing. (If other moral demands, such as ensuring public safety, justify disclosing proprietary information in a specific situation, the disclosing party should be able to produce that justification.) Obeying moral rules, fulfilling obligations, and respecting others' rights may require conscientiousness, even courage, but seldom require creative thinking.

Statements of prospective responsibility, which specify the ends to be achieved rather than the acts to be performed, require more creative reasoning. For example, fulfillment of an engineer's responsibility for safety requires understanding the safety hazards posed by a given situation or technology and figuring out the best way to reduce or eliminate those hazards. An engineer's responsibility to promote the public understanding of technology requires knowledge of the technology, an assessment of (some segment of) the public's understanding of it, a knowledge of the opportunities available for improving that understanding, and finally the construction of a statement or presentation that fits the situation and is appropriate to the current level of public understanding.

Statements of prospective responsibility do not provide directions for doing what needs to be done or in what order they should be done. The synthetic tasks of devising appropriate actions are usually not completed before action is taken; they are continually revised in light of changing circumstances and new discoveries. The exercise of creative or synthetic reasoning in fulfillment of professional responsibility does not necessarily require originality (a *novel* synthesis); it only requires a synthesis appropriate to the situation. However, an appropriate synthesis can be extremely challenging for the kinds of multiply constrained problems engineers typically face. I agree with Woodie Flowers that creation under multiple constraints is *more* challenging than artistic creation, which is typically less constrained (Woodie C. Flowers, Pappalardo Professor of Mechanical Engineering, Massachusetts Institute of Technology, personal communication).

In some borderline cases, of course, obligation may shade into responsibility. For example, maintaining a client's confidentiality might require expert judgment to ensure that information is not disclosed; in that event, the requirement to ensure confidentiality is a borderline case.

Significant moral problems, such as how best to fulfill one's responsibilities, are like design problems, especially engineering design or experimental design problems (Whitbeck, 1998). Both sorts of problems require synthesis as well as analysis. Moral problems are *not* multiple-choice problems, "decision problems" (in the technical sense), or "dilemmas" (*literally,* multiple-choice problems in which all of the choices are unacceptable). In other words, they are not problems that require choosing between *preexisting alternatives.* Although it may be useful to assign multiple-choice ethical problems to teach certain lessons, understanding the differences between the structure of actual moral problems and multiple-choice problems (including dilemmas) is important for developing the full range of skills necessary to moral reasoning and moral problem solving.

Here are some common features of interesting or substantive engineering design problems and moral problems:

- There is rarely, if ever, a uniquely correct solution. If there is any solution, there is usually more than one.
- Although there is no uniquely correct solution, some possible responses are clearly unacceptable; there are wrong answers even if there is no unique right answer, and some solutions are better than others.
- Two solutions may have different advantages. Therefore, it is not necessarily true that, for any two candidate solutions, one must be incontrovertibly better than the other.
- Any acceptable solution must do the following things:
  1. Achieve the desired end (e.g., design the requested item or fulfill one's responsibility).
  2. Conform to given specifications or explicit criteria for this act (e.g., for the designed item: meet size requirements; for the responsibility: not require an inordinate amount of time that causes one to forego other major responsibilities).
  3. Be reasonably secure against accidents and other mishaps.
  4. Be consistent with background constraints that are often unstated (e.g., a consumer item should be affordable and not use very hazardous materials; the response to an ethical problem should not violate anyone's human rights).

The analogy between moral problems and design problems draws attention to several frequently neglected features of moral problems. First, morally relevant considerations, analogous to design constraints, should not be assumed to be opposed to each other. Second, satisfying one moral demand does not generally mean disregarding others. I emphasize this point because inexperienced teachers of professional ethics often simplify moral problems and present them as choices between two values—for example, between loyalty to one's employer and devotion to public safety or between policies that protect the environment and policies that further job growth. Simplifying moral problems encourages stereotypic thinking, rather than critical thinking, and so closes students' minds to the possibility of satisfying multiple moral demands simultaneously. For example, one can further the welfare of one's client or employer *by* preventing a disastrous accident. Designers consider many design criteria simultaneously, and teachers of engineering ethics must foster similar skills.

LIMITS OF FORESIGHT AND RESPONSIBILITY

My fifth topic is limitations on the foresight necessary for the exercise of responsibility. Here I make contact with the themes of other papers in this volume: specifically, the concerns raised by Wm. A. Wulf about complexity and unpredictability (see pp. 1–6 in this volume) and Braden Allenby's discussion about the macro-effects of human action (see pp. 9–27 in this volume). The scope of engineering responsibility (what some have called the "problem space" of engineering) has expanded repeatedly in the course of the twentieth century as engineers have been called upon to consider a greater range of factors and to foresee a wider range of consequences. We must remember that there are limits to what engineers can foresee, and hence how effectively they, individually or in teams, can achieve ethical ends, such as safety of the public.

Henri Petroski (1985) has argued that engineering often advances by learning from failures and accidents and that those experiences have broadened the range of factors engineers must consider in fulfilling their responsibilities. Experience with the consequences of engineering design decisions has widened the scope of consequences responsible engineers are expected to foresee and the range of factors they are expected to consider in controlling those consequences. Not only the number of factors, but also the range of eventualities has increased. For example, automobiles are not intended to have collisions, but they can be expected to have them. The goal of reducing injuries and damage from automobile accidents is, therefore, now recognized as a responsibility of automotive designers.

The list of questions below illustrates how much the scope of engineering considerations has expanded in the crucial area of safety. (The responsibility for safety might be replaced with other responsibilities in engineering practice or engineering research or the ethical treatment of human and animal subjects.) The responsibility to ensure that a device or construction is safe in its intended use is only the beginning of what engineers must consider to fulfill that responsibility:

- Will the device or construction operate safely under conditions for its intended use? Example: boiler explosions.
- Will the device or construction be safe in accidents that are likely to occur? Example: boating accidents.
- Will the device or construction be safe under condition of common misuse? Example: children playing "house" in clothes dryers. (Attempting to forestall *every possible* harmful misuse may be self-defeating, as well as

paternalistic, inasmuch as one may thereby block important beneficial uses that would have a net positive effect on health and safety.)

- Will the device or construct be safe if maintained in a way that may be improper but is likely to be a temptation given the design? Example: the 1979 American Airline DC-10 crash caused by cracks in the flange of the rear bulkhead resulting from time-saving shortcuts in maintenance procedures.
- Will the device or construct be safe in interactions with other technologies? Example: a patient's death that showed the need for an electrical ground isolation standard in medical devices. A patient who had survived a heart attack lay in his hospital bed with an electrocardiograph attached to his chest and plugged into the wall. He also had an internal heart-pressure catheter, which was plugged into the opposite wall. In the next room a janitor was operating a vacuum cleaner that had a near short. When the vacuum cleaner was plugged into the wall, it caused current flow in the ground wire, killing the patient (Woodie C. Flowers, personal communication).

As we try to anticipate the problems engineers will face in the twenty-first century, we must consider not only how they will handle responsibilities in designing, testing, manufacturing, and recycling new technologies that raise considerations similar to those raised by previous technologies, but also how they will handle an expanded range of considerations. Another major question is under what circumstances the possible effects of accidents might be so great that we cannot afford to "learn from experience."

Normal Accidents in Complex Systems

Although expanding the scope of design criteria based on lessons learned from failures and accidents has made the design of devices and components safer, these lessons may be of little use in addressing what Charles Perrow has called "normal accidents" to which technologically sophisticated, complex systems fall prey. Perrow (1984) coined the term "normal" or "system" accidents to describe accidents with the following characteristics:

- They arise in "tightly coupled" (time-constrained) complex systems.
- They involve the unanticipated interaction of multiple failures of components.
- They involve an interaction of component failures that neither the operator *nor the designer* could anticipate or comprehend.
- They are, nonetheless, often attributed to operator error.

Because normal accidents arise from complexity, they cannot be remedied with technical fixes, such as safety devices. In very complex systems, a safety device often creates another component subject to failure—which failure could interact with other failures to produce situations that defy ready diagnosis. In addition, some safety devices may allow for more risky behavior—for example, safety devices for marine transport have allowed captains, spurred by competition, to increase their speed, so there has been no reduction in the accident rate.

As the designer's adage reminds us, "the best part is no part at all." So, perhaps, redesign for safety may be accomplished through reduction in complexity, rather than the addition of safety devices. Furthermore, as Michael Loui reminds us, ARPANET, a predecessor of the Internet, was specifically designed to withstand various kinds of failures (such as lost packets, noisy communication links, and failed nodes). Thus, in some cases, engineers have had great success in designing against failure. However, reducing complexity often requires broader collaboration and may not be within the control of engineering designers or design teams.

New Technologies

Since the late 1970s, Michael Martin and Roland Schinzinger (1983) have argued that technological innovation amounts to social experimentation and, therefore, requires informed consent analogous to the consent from patients for using experimental therapies. Note that the informed consent for use of an experimental therapy is quite different from the consent for human subjects in experimental studies. Martin and Schinzinger's analogy is not between engineering innovation and clinical experimentation, but between engineering innovation and the use of experimental medical treatment. The purpose of technological innovation, like medical therapy, is to meet a practical need, not simply to acquire knowledge. The use of an experimental medical treatment is governed by standards of competent care and informed consent for care, rather than the more stringent norms applied to clinical experiments.

Because experimental therapies may pose significant unanticipated risks to health and safety or to other major aspects of well-being (such as financial security) that patients are best able to appreciate, it is widely agreed that patient consent should be obtained before such therapies are used. However, informed choice generally requires help from experts—for example, physicians must outline the possible risks and benefits and the therapeutic alternatives.

Estimating the possible social, economic, political, and environmental consequences of developing or adopting a new technology also generally requires expert knowledge, often from engineering experts *and* experts in other disciplines. Therefore, to fulfill their responsibility for educating the public about a new technology, engineers will have to collaborate with other disciplines. Martin and Schinzinger proposed using "proxy groups" composed of people similar to those who would be greatly affected by a new technology to assess the extent of possible harm or benefits. (I understand that experiments with citizen panels are now being conducted.) The challenge is to describe the consequences of new technologies or new uses of technologies in a clear and convincing way. The consequences will presumably include not only health and safety risks, but also social, economic, political, and environmental risks. Engineering expertise will certainly be necessary to characterize many risks and consequences, but it will not be sufficient for characterizing all of them. Developing the interdisciplinary collaboration for estimating consequences in a rapidly evolving social and technological environment will present serious challenges.

RESPONSIBILITY, CREATIVITY, AND CHANGE

Finally, we must consider how engineering knowledge can be brought to bear on societal choices about technology in an age of technological complexity and rapid social and technological change. We have seen that technological innovation requires not only addressing multiple, sometimes competing, design constraints, but also extending foresight into new areas. For example, design considerations have been expanded to include how a design might provide incentives for improper and/or unsafe maintenance procedures.

We live in a time of rapid social change, as well as technological innovation, and social change makes the consequences of cumulative technological change more difficult to foresee and predict. The interaction of medical technologies with a short in janitorial equipment illustrated the interaction of technologies. Social change, particularly when a significant number of people begin to use technologies for new purposes, can increase risks significantly. For example, there has recently been an increase in the risk of sabotage, first by computer hackers and more recently by terrorists. Rapid change makes it difficult to use prior engineering experience to guide current practice.

In addition, we are confronted with the unpredictability of complex systems that Bill Wulf discussed in his paper for this workshop. Thus, we are left with two distinct, crucially important questions:

- What is the best way to prepare engineers to fulfill their responsibilities for consequences they can, *in principle*, foresee?
- What is the best way to integrate engineering expertise with non-engineering knowledge (both lay and expert) to define the scope and limits of the problems engineers are now being asked to solve?

Answering the second question will require determining on which problems engineering knowledge should be brought to bear, which risks are too great to allow a technology to be pursued, and how we can reduce the complexity that gives rise to normal accidents and inherently complex systems. Such delimitation *cannot* be accomplished by engineering design alone. Thus, answering the second question will require creative—and interdisciplinary—skills on the part of engineers.

REFERENCES

Ladd, J. 1970. Morality and the ideal of rationality in formal organizations. The Monist 54(4): 488–516.

Loui, M.C. 1998. The engineer's responsibility for quality. Science and Engineering Ethics 4(3): 347–350.

Martin, M.W., and R. Schinzinger. 1983. Ethics in Engineering. New York: McGraw-Hill.

Perrow, C. 1984. Normal Accidents. New York: Basic Books.

Petroski, H. 1985. To Engineer Is Human. New York: St. Martin's Press.

Whitbeck, C. 1998. Ethics in Engineering Practice and Research. Cambridge, U.K.: Cambridge University Press.

Microethics, Macroethics, and Professional Engineering Societies

JOSEPH R. HERKERT
North Carolina State University

Engineering ethics can be considered in three frames of reference—individual, professional, and social—which can be further divided into "microethics" (concerned with individuals and the internal relations of the engineering profession) and "macroethics" (concerned with the collective, social responsibility of the engineering profession and societal decisions about technology). Research and instruction in engineering ethics have traditionally focused on microethical issues and problems, and little attention has been paid to macroethics or the integration of microethical and macroethical approaches. In this paper I define and explain the importance of considering both microethics and macroethics and discuss (1) how professional engineering societies can link individual and professional ethics and (2) how they can link professional and social ethics.

MICROETHICS AND MACROETHICS IN ENGINEERING

The political philosopher Langdon Winner (1990) has criticized the overemphasis in engineering ethics on case studies of individual dilemmas and the neglect of more global issues related to the development of technology:

> Ethical responsibility . . . involves more than leading a decent, honest, truthful life, as important as such lives certainly remain. And it involves something much more than making wise choices when such choices suddenly, unexpectedly present themselves. Our moral obligations must . . . include a willingness to engage others in the difficult work of defining what the crucial choices are that confront technological society and how intelligently to confront them.

TABLE 1 Some Microethical and Macroethical Issues in Science and Engineering

	Engineering Practice	Scientific Research
Microethics	Health and safety	Integrity
	Bribes and gifts	Fair credit
Macroethics	Sustainable development	Human cloning
	Product liability	Nanoscience

Indeed, in the past 30 years as engineering ethics has emerged as an academic subfield, several authors, including ethicist John Ladd (1980), have issued similar critiques, noting that engineering ethics must encompass multiple domains (Herkert, 2001). This is also true of ethics in many other fields, such as ethics in research (Table 1).

One way of expanding engineering ethics to address macroethical issues is to consider the ethical implications of public policy issues, such as risk and product liability, sustainable development, health care, and information and communication technology (Herkert, 2000). Although, the melding of ethics and professionalism has significantly contributed to the development of concepts in engineering ethics, the emphasis to date has been on issues *internal* to the profession, giving short shrift to macroethical issues (O'Connell and Herkert, 2004). There are some indications, however, that a more balanced view is gradually taking hold.

ROLE OF PROFESSIONAL SOCIETIES

The distinction between microethics and macroethics is useful for mapping the role of professional societies in engineering ethics (Herkert, 2001). So far, the role of professional engineering societies has been limited largely to developing codes of ethics. Professional societies, however, could potentially serve as a conduit to bring together the entire continuum of ethical frameworks by linking individual and professional ethics and linking professional and social ethics. In the domain of macroethics, professional societies can provide a link between the social responsibilities of the profession and societal decisions about technology by issuing position statements on public policy issues, such as sustainable development (Herkert, 1998) and product liability reform (Herkert, 2003a).

In the microethical domain, professional societies can provide support for individuals who engage in ethical behavior. Engineers and other professionals who blow the whistle on unethical behavior or otherwise take action consistent with their code of ethics often pay a heavy price, which may include demotion, firing, blacklisting, or even a threat to life. Under these circumstances, many have argued that it is unreasonable to expect individual engineers to be "moral heroes."

Scholars have focused a great deal of attention on how professional societies can provide support for engineers who act ethically, on the grounds that members of a society have a collective responsibility to promote and protect ethical behavior (Ladd, 1982). When efforts to provide ethics support through corporate ethics offices and government regulation meet with mixed results (Herkert, 2000), professional engineering societies can provide a counterweight to the pressures of the workplace (Unger, 1994).

PROFESSIONAL SOCIETIES AND ETHICS SUPPORT

Codes of engineering ethics give primacy to public safety, health, and welfare, thus implying that they support individual engineers whose actions are consistent with these goals and other provisions of the codes (Herkert, 2001). Unfortunately, evidence suggests that professional societies have an uneven history of providing ethics support (Herkert, 2001, 2003b). In fact, they often seem unwilling or unable to provide sustained support for the ethical behavior of their members. Take, for example, the recent record of the Institute of Electrical and Electronics Engineers (IEEE) wherein gains in ethics support that had been long sought after were crushed by a backlash by staff and volunteers. This example is all the more striking because IEEE is often regarded as one of the most progressive professional societies in the ethics arena (Unger, 1994).

Like other codes of engineering ethics, the *IEEE Code of Ethics*, implemented in 1990, pledges its members "to accept responsibility in making engineering decisions consistent with the safety, health and welfare of the public, and to disclose promptly factors that might endanger the public or the environment." Unlike some other codes, however, the IEEE code also includes specific language regarding ethics support, committing its members "to assist colleagues and co-workers in their professional development *and to support them in following this code of ethics*" [emphasis added] (IEEE, 1990).

Prior to 1995, the only committee at the level of the IEEE Board of Directors charged with dealing with ethics was the Member Conduct Committee (MCC), established in 1978, whose purpose was two-fold: (1) to recommend disciplinary action for members found to be acting in violation of the code of ethics; and (2) to recommend support for members who, in following the code of ethics, have been retaliated against (Unger, 1999). A board-level Ethics Committee, formed in 1995 as a result of efforts by members (including members of the IEEE subunit that represents U.S. members) to elevate the status of ethics in IEEE, was intended to keep members informed and advise the Board of Directors on ethics-related policies and concerns.

In 1996, one of the first actions taken by the Ethics Committee was to establish an "ethics hotline" to provide information and advice on ethical matters to professionals in IEEE's field of interest. Cases brought to the attention of the hotline included falsification of quality tests, violations of intellectual property rights, and design and testing flaws that could compromise public safety. Some of these cases were referred to and acted upon by the MCC (Unger, 1999).

The Executive Committee of the Board of Directors suspended the IEEE ethics hotline in 1997 after less than a year of operation (Unger, 1999). In 1998, the Executive Committee rejected and suppressed its own task force report that recommended reactivation of the hotline. In the same year, IEEE implemented changes in its by-laws that shortened the terms in office of members of the MCC and Ethics Committee and, in apparent disregard of IEEE's own code of ethics, prohibited the Ethics Committee from offering advice to individuals, including IEEE members. The cycle was completed in 2001 when the Ethics Committee and MCC were merged. Like the old MCC, the combined committee has a dual charge of ensuring member discipline and providing ethics support, but its activities are limited by IEEE by-law I-306.6, which provides, "Neither the Ethics and Member Conduct Committee nor any of its members shall solicit or otherwise invite complaints, nor shall they provide advice to individuals" (IEEE, 2001). Nevertheless, the provision of the *IEEE Code of Ethics* calling for ethics support has not been changed.

Opponents of ethics support often cite liability concerns as a rationale, an argument that Unger has refuted persuasively (Unger, 1994, 1999). In addition, some are concerned that an ethics hotline would put IEEE in the undesirable position of mediating disputes between members and their employers. In other words, corporate influence is a factor in the reluctance of professional societies to provide ethics support.

Layton (1986), for example, describes engineers as part scientists and part businesspersons, but not really either; he says they are marginal in both contexts. This situation, the result of the concurrent development of engineering as a profession and technology-driven corporations, inevitably leads to conflicts between the professional values of engineering and business values. Layton notes that professionals value autonomy, collegial control, and social responsibility, while businesses emphasize loyalty, conformity, and the overarching goal of improving the bottom line. This tension is exacerbated when the career paths of engineers lead to management positions. Engineers who hope to advance in the corporate hierarchy are expected to embrace business values.

Davis (1998) disputes the argument that there is an inherent conflict between an engineer's status as an employee and his or her professional autonomy. As Layton points out, however, many of the leaders of professional societies are senior members who have moved from technical engineering into business management positions. In addition, many companies encourage and fund the participation of their employees in professional societies.

Another possible explanation for the reluctance of professional societies to provide ethics support is that the engineering/business culture places a high premium on economic efficiency and downplays the societal context of engineering. "The engineering view" is often characterized as focusing mainly on technical solutions to problems, which may account for the unwillingness or inability of some to recognize the social and ethical dimensions of engineering (Herkert, 2000). Other factors that may contribute to the reluctance to provide ethics support include an unwillingness to air the profession's dirty laundry in public and perceived complications related to the increasing globalization of professional societies.

PROFESSIONAL SOCIETIES AND PUBLIC POLICY

Although professional engineering societies also have a mixed record in advancing ethical principles in the macro-arena (Herkert, 2003a), there have been some hopeful signs recently, notably in the case of sustainable development, which has become a major public policy issue worldwide, including in the engineering and business communities (Herkert, 1998). Following the publication of the Brundtland Commission report in 1987, which defined sustainable development as "development that meets the needs of the present without compromising the ability of future generations to meet their own needs," the concept attracted considerable attention in the international community and on national agendas (WCED, 1987). In 1992, the United Nations Conference on Environment and Development in Rio de Janeiro issued *Agenda 21*, a blueprint for global sustainable development that led to the establishment of dozens of national commissions, including the President's Council on Sustainable Development in the United States (Agenda 21, 1992).

The engineering community reacted to *Agenda 21* by establishing the World Engineering Partnership for Sustainable Development (WEPSD) in 1992 (Carroll, 1993); and committees formed by the traditional engineering organizations, including the American Association of Engineering Societies and discipline-based societies, such as the American Society of Civil Engineers (ASCE), issued position papers.

The theory of sustainable development, which emerged from the field of ecological economics, involves achieving objectives in the ecological, economic, and social realms. The ecological objective is to maintain a sustainable scale of energy and material flows through the environment that does not erode the carrying capacity of the biosphere. The economic objective is to allocate resources efficiently in conformance with consumer preferences and the ability to pay. The social objective is to distribute resources justly among people, including future generations. The overall objective is sustainability in economic, ecological, and social systems (Farrell, 1996).

An alternative way of characterizing development is to think of three distinct systems—biological, economic, and social—each of which has its own goals. Sustainable development is achieved when, after the inevitable trade-offs and setting of priorities for a given time or place, these goals are maximized in all three systems. The International Institute for Environment and Development lists typical goals for each system (Holmberg and Sandbrook, 1992):

Biological (ecological) system
- genetic diversity
- resilience
- biological productivity

Economic system
- increased production of goods and services
- satisfaction of basic needs or reducing poverty
- improvements in equity

Social system
- cultural diversity
- social justice
- gender equality
- participation

Although it is still a controversial concept, sustainable development maintains considerable currency in a number of circles, including engineering. Some engineering societies have even proclaimed sustainable development to be an ethical responsibility (Grant, 1995).

The success of public policy promoting sustainable development depends upon achieving all of the objectives of a sustainable society. However, despite proclamations that engineers have an ethical responsibility to promote sustainable development, questions about just distribution and other aspects of equity (such as risk distribution) were often excluded when engineers first began to consider policies and issues (Herkert, 1998). Indeed, engineering organizations focused almost exclusively on striking a balance between economic development and environmental protection. Although their efforts were commendable, they were limited by their failure to come to grips with the third essential element of sustainable development—the social objective. In early statements, it appeared that engineers either were not interested in or were not able to articulate social concerns.

Despite this shortcoming, some engineers argued that engineering should be accorded a preeminent position, thus endorsing a technocratic vision of sustainable development (Herkert, 1998). A founder of WEPSD went so far as to portray engineers as the best arbiters of *all* knowledge that must be brought to bear on the problem (Carroll, 1993).

Recently, however, some engineering societies have included the social objective in the role of engineering in the realization of sustainable development.

This can be seen clearly in a document prepared by several U.S.-based engineering societies for the Johannesburg Earth Summit in 2002 (ASME, 2002):

> Creating a sustainable world that provides a safe, secure, healthy life for all peoples is a priority for the US engineering community. It is evident that US engineering must increase its focus on sharing and disseminating information, knowledge and technology that provides access to minerals, materials, energy, water, food and public health while addressing basic human needs. Engineers must deliver solutions that are technically viable, commercially feasible, and environmentally *and socially* sustainable [emphasis added].

The willingness to acknowledge social sustainability reflects a maturity of thought and sensitivity to societal and ethical issues rarely found in the deliberations of professional societies on microethical issues (or on many macroethical issues).

We may ask why professional societies are (sometimes) willing to advocate ethically sensitive public policies, whereas they have typically been timid about addressing ethical controversies involving individuals. I propose three preliminary explanations (Herkert, 2003b):

- Macroethical issues are well suited to cooperative action among many professional engineering societies. Collective action can often offset corporate influences, and even transcend international boundaries.
- The leaders of professional societies can be agents of change in the engineering culture (if they choose). Because macroethical issues affect all members of the profession, they are ideal vehicles for promoting change.
- Responding ethically to macroethical challenges provides an opportunity for professional societies to improve the public image of engineering as a *by-product* of ethical action rather than as the *goal* of ethical posturing.

REFERENCES

Agenda 21. 1992. New York: United Nations Publications. Available online at *http://www.un.org/ esa/sustdev/documents/agenda21/english/agenda21toc.htm.*

American Society of Mechanical Engineers (ASME). 2002. A Declaration by the U.S. Engineering Community to the World Summit on Sustainable Development. ASME position statement. Available online at: http://www.asme.org/gric/ps/2002/02-30.html.

Carroll, W.J. 1993. World engineering partnership for sustainable development. Journal of Professional Issues in Engineering Education and Practice 119: 238–240.

Davis, M. 1998. Thinking Like an Engineer. New York: Oxford University Press.

Farrell, A. 1996. Sustainability and the design of knowledge tools. IEEE Technology and Society 15(4): 11–20.

Grant, A.A. 1995. The ethics of sustainability: an engineering perspective. Renewable Resources Journal 13(1): 23–25.

Herkert, J.R. 1998. Sustainable development, engineering and multinational corporations: ethical and public policy implications. Science and Engineering Ethics 4(3): 333–346.

——. 2000. Social, Ethical and Policy Implications of Engineering. New York: IEEE Press.

——. 2001. Future directions in engineering ethics research: microethics, macroethics and the role of professional societies. Science and Engineering Ethics 7(3): 403–414.

——. 2003a. Professional societies, microethics, and macroethics: product liability as an ethical issue in engineering design. International Journal of Engineering Education 19(1): 163–167.

——. 2003b. Biting the apple (but not inhaling): lessons from engineering ethics for alternative dispute resolution ethics. Penn State Law Review 108: 119–136.

Holmberg, J., and R. Sandbrook. 1992. Sustainable Development: What Is to Be Done? Pp. 19–38 in Making Development Sustainable, edited by J. Holmberg. Washington, D.C.: Island Press.

IEEE (Institute of Electrical and Electronics Engineers). 1990. IEEE Code of Ethics. Available online at *http://www.ieee.org/about/whatis/code.html.*

IEEE. 2001. By-law I-306. Available online at *http://www.ieee.org/about/whatis/bylaws/i-306.html.*

Ladd, J. 1980. The Quest for a Code of Professional Ethics: An Intellectual and Moral Confusion. Pp. 154–159 in AAAS Professional Ethics Project: Professional Ethics Activities in the Scientific and Engineering Societies, edited by R. Chalk, M.S. Frankel, and S.B. Chafer. Washington, D.C.: American Association for the Advancement of Science.

Ladd, J. 1982. Collective and individual moral responsibility in engineering: some questions. IEEE Technology and Society Magazine 1(2): 3–10.

Layton, E.T. 1986. The Revolt of the Engineers. Baltimore, Md.: Johns Hopkins University Press.

O'Connell, B., and J.R. Herkert. 2004. Engineering ethics and computer ethics: twins separated at birth? Techné: Research in Philosophy and Technology 8.

Unger, S. 1994. Controlling Technology: Ethics and the Responsible Engineer, 2nd ed. New York: John Wiley and Sons.

Unger, S. 1999. The assault on IEEE ethics support. IEEE Technology and Society Magazine 18(1): 36–40.

WCED (World Commission on Environment and Development) (The Brundtland Report). 1987. Our Common Future. Oxford, U.K.: Oxford University Press.

Winner, L. 1990. Engineering Ethics and Political Imagination. Pp. 53–64 in Broad and Narrow Interpretations of Philosophy of Technology, edited by P. Durbin. Philosophy and Technology 7. Boston: Kluwer.

Ethics in Engineering Education

Ethics across the Curriculum
PREPARING ENGINEERING AND SCIENCE FACULTY TO INTRODUCE ETHICS INTO THEIR TEACHING

VIVIAN WEIL
Illinois Institute of Technology

Emerging technologies have been in the forefront of attention since I began teaching and doing research on engineering ethics in late 1976. At that time, the subject was nuclear power and engineers' responsibilities in designing, maintaining, and regulating nuclear power plants. Later, our attention was captured by ethics and responsibility in agricultural biotechnology, even though this field does not clearly count as an engineering specialty. Today, emerging uses of information technologies generate ethical issues, for example, protecting human subjects in online research. Ethical issues generated by burgeoning developments in nanoscience and nanotechnology are just coming into view.

But my direct concern here is not with emerging technologies. My focus is on preparing engineering and science faculty to introduce ethics into their teaching. An important aim of teaching ethics is to prepare engineering students to identify and cope responsibly with ethical issues in emerging technologies.

My plan is to describe the Ethics across the Curriculum Workshops, which are designed to prepare faculty to introduce ethics teaching into their regular courses. My colleague Michael Davis has conducted this program for engineering and science faculty at Illinois Institute of Technology (IIT) since 1991. Made possible by funding from the National Science Foundation (NSF), the workshops were offered at first only to IIT faculty, but later they were also offered to faculty from other universities. In the 2002 and 2003 workshops, several faculty members from overseas were among the participants.

HISTORY OF THE PROJECT

The Ethics across the Curriculum Project was started in the late 1980s when two young, energetic research faculty members came to the Ethics Center concerned that something was missing from their teaching. They felt they should include more of the context and complexities of actual engineering problems and in so doing bring out their ethical aspects. They identified ethical issues associated with some of the topics in their courses they thought should be raised, but they did not know how to address these issues in their teaching.

In response to this call for help, I organized sack-lunch meetings of interested faculty to discuss options for addressing these concerns. After coming together regularly over a considerable period of time, faculty members agreed on the importance of teaching ethics in engineering. They also agreed that what kept them from teaching ethics was a lack of necessary skills and experience. In addition, they felt that teaching ethics would not be legitimate because it was not part of their graduate training. Yes, they knew something about ethics as members of society, but they knew about many things they did not feel prepared to teach.

With this insight, we set out to develop a program, and, working together, we devised a workshop plan. During our discussions, we had noted that in times past engineering educators had favored a diffusion method of teaching ethics. However, we had never seen a plan for a diffusion method that specified what faculty should do in their classrooms, how their teaching of ethics would be evaluated, or how student evaluations would feed back into the program, let alone how a program would be monitored to make sure that diffusion teaching was taking place.

We knew of some precedents, including workshops that had been tried at several other universities and one workshop on ethics for business educators at the old Arthur Anderson Company campus in St. Charles, Illinois. Reports by participants in those exercises indicated that they felt that they had learned a good deal—or at least had found the programs interesting. But they did not see how to connect what they had learned in the workshops with their teaching. Therefore, we thought it essential to adopt a nuts-and-bolts approach, that is, an approach likely to help faculty actually begin teaching ethics.

Our sack-lunch discussions eventually led to a proposal to NSF. It was gratifying to note that the first paragraphs of the proposal were written by one of the two colleagues who had approached us originally and prompted us to undertake our discussions. NSF funded the first proposal for four years, the first three years limited to IIT faculty and the last year for faculty from other institutions. Subsequent funding from NSF made it possible for us to conduct workshops for faculty from other institutions almost every year until the last workshop in the summer of 2003.

The funding covered not only the operation of the workshops and stipends for the instructors, but also stipends for participants matched by stipends from the participants' own institutions. The rationale for providing stipends for the 15 to 20 participants each year was to attract very able, busy people who had other interests competing for their time. The stipends underlined the importance of ethics and the honor of being accepted in the program.

UNDERLYING ASSUMPTIONS

Assumptions underlying the workshops were made explicit. First, ethics is not peripheral to, or an add-on to, engineering. It is integral to the practice of engineering, part of engineering problem solving. Safety and guarding against avoidable harm are built into engineering; they are the principles that underlie engineering codes and standards.

Second, engineering faculty should be engaged in the teaching of ethics. They not only have more exposure to students than ethics specialists and others in the humanities, but they also have the credibility to convey the importance of ethical considerations in problem solving. For many faculty members, learning to teach ethics is a feasible undertaking, provided they start with modest changes after suitable preparation.

For example, they might start with a problem in the back of the book that can be fleshed out, such as a problem concerning the flow of fluids. To make the problem less abstract, the fluids can be described as flowing into a reservoir. Providing information about the destination of the fluids and the nature of the fluids supplies context that is often absent from the problems students work on. Concrete details help bring ethical questions to the surface. This problem, for example, raises questions about whether the fluids are acceptable in a reservoir for drinking water.

I want to emphasize that we are talking about small changes. We believe that people can begin very modestly, and as they gain more confidence, more familiarity with the materials they can use, and more skills, they can do more. Explicit, thoughtful consideration of relevant ethical questions in a homework problem can be enough to engage students' interest and open the way for a continuing focus on ethics.

Third, ethics material is a normal component of the course. This means that students should be held responsible for mastering this material, as they are for mastering other components of the course—through grading.

THE PLAN

Each seven-day workshop included five days of lectures and discussions. The instructors played a prominent role in the first few days, but as the workshops proceeded, participants gradually moved to the fore. Participants some years back introduced an innovation, a role-play of a faculty senate meeting, that has become an important component of the fifth workshop day. The lectures throughout the workshop were relatively short, and considerable time was allowed for discussion. The sixth and seventh days were usually held after a two-day interval over a weekend. On the sixth day, faculty participants did most of the work, presenting material they planned to use in their teaching, often a problem for homework or a class problem. After each presentation—an explanation of how the speaker planned to use the problem or assignment—the audience, including the instructor, raised questions and offered criticisms. Then the speaker responded. The atmosphere was much like that of a graduate seminar. On the seventh day, the participants brought in other examples of material they planned to use in teaching with explanations of how they planned to grade students' responses. Again other participants and an instructor offered questions and comments, and the speaker responded.

In advance of the workshops, participants received assigned readings of selected articles and one book. The book, now in its fourth edition, *The Elements of Moral Philosophy* by James Rachels (McGraw Hill, 2002), turned out to be very useful. This slim volume is philosophically sound and covers the leading theories. Faculty participants seemed to find it interesting and readable.

Each workshop had two instructors. Michael Davis, the principal investigator and author of *Thinking Like an Engineer* (Oxford University Press, 1998), planned the program and served as lead instructor. Robert Ladenson, professor of philosophy in the Lewis Department of Humanities at IIT, and I alternated from year to year as the second instructor.

WORKSHOP: DAY ONE

Getting acquainted was the focus at the start of the first day. In the first lecture, Davis offered definitions of key concepts—prudence, morality, law, and ethics—to clarify major concepts in practical and professional ethics. For example, he defined morality as the standards everybody wants everybody else to conform to, so much so, that each of us is willing to follow those standards ourselves. In the discussion that followed the lecture, participants were invited to raise questions and offer counterexamples. The aim was to encourage careful use of familiar concepts and to convey to participants how much they already knew about ethics.

The second part of the morning was devoted to a well known case, "Catalyst B," which was first published in a chemical engineering magazine in 1980. The discussion went forward without direction or guidelines. Participants simply read the case and began to talk about it in an unstructured discussion resembling a rap session.

The unguided discussion was followed by a presentation on method—what to look for, how to argue, and guidelines for discussion— a canonical set of seven steps (some variations include only five steps). These steps are essentially a checklist for orderly discussion that drives to a conclusion, so that in the end one makes a judgment.

The presentation was followed by a second discussion of the case, this time against the background of the guidelines and some acquaintance with ways to argue. The second discussion was more controlled and orderly and led to a conclusion.

WORKSHOP: DAY TWO

On the second day, the focus was on moral theory. The aim was to acquaint the participants with various philosophical ethical theories. Participants could recognize the features of morality, such as consequences or duties, emphasized in a particular theory. They came to understand that, although theories cannot be used mechanically to analyze issues and resolve cases, exposure to the theoretical tradition in ethics is valuable, if only because it enables one to recognize when a student's probing questions reveal an interest in theory.

We began with consequentialist theories, specifically utilitarianism. After a brief lecture describing the theory, we discussed a case ("New Software Case," a short film produced by the Ethics Resource Center) focusing on utilitarian reasoning. This was followed by a short lecture on Kant's rule-based ethical theory, with particular attention to some Kantian ideas, such as respect for persons. Respecting the inherent dignity of every human being means that people can never be "used" merely as means to an end. The discussion that followed highlighted Kantian reasoning in the "New Software Case."

Finally, there was a short lecture covering a group of ethical theories: social justice; virtue theories; pluralism; relativism; and moral minimum. Another discussion of the same case brought out examples of reasoning according to each theory.

WORKSHOP: DAY THREE

The third day was focused on professionalism and professional ethics and introduced codes of ethics. Engineers have been especially active in producing both technical standards and ethical standards that make explicit the values and principles that underlie the technical standards. On this day, the group reconsidered the case from Day One, "Catalyst B," this time with the emphasis on professionalism and the use of professional ethics codes in problem solving.

A brief lecture on context of professional work and strategies for identifying issues provided a bridge to a presentation on teaching ethics in the classroom. This session ended with a general discussion of various techniques to use in teaching: the case method; vignettes; role playing; debates; and "Ethics Bowl."

"Ethics Bowl," an engaging competition inspired by *College Bowl* and devised by Robert Ladenson at IIT, features open-ended questions about ethics scenarios. "Ethics Bowl" has grown into a national competition for 40 institutions at the annual meeting of the Association for Practical and Professional Ethics.

WORKSHOP: DAY FOUR

Day four was devoted to both cognitive questions and moral questions about teaching ethics. For instance, some have questioned whether virtue can be taught, an old question first discussed by Plato. There is now a considerable literature providing good evidence that cognitive moral learning goes on in higher education, at the undergraduate, graduate, and professional levels. Moral learning is just part of the learning that goes on in institutions of higher education.

Participants also raised concerns about the ethics of teaching ethics. The risk of indoctrinating students was also a common concern. Clearly, teaching ethics should not become an occasion for promoting one's own views. Yet there is a wide range of opinion about whether to what extent faculty are justified in injecting or revealing their own positions. At the same time, there is wide agreement that the classroom environment should make students feel comfortable about airing their own views. Allowing a variety of opinions to come out can give students an opportunity to note differences in assumptions and conclusions. That experience can prompt students to think through and assess their own positions.

It is essential to make explicit the goals of teaching any subject. Articulating the goals of teaching ethics can be helpful in responding to cognitive and moral concerns. As teachers in universities we aim to lead students to think. In teaching ethics, our goal is to stimulate thinking about ethical issues and to help students acquire analytical skills that will be useful to them as they consider ethical issues in their professional lives. The experience of sitting in a classroom with other students and an instructor intent upon resolving ethical problems may strengthen their resolve to behave responsibly. And strengthening that resolve is a legitimate goal of ethics teaching.

WORKSHOP: DAY FIVE

The fifth day was focused on pedagogy. Beginning in the second year of the workshops, the first half of the fifth morning was devoted to reports by two faculty members who had been through a workshop. Their accounts of their experiences—what they would do again, what mistakes they made—were helpful and encouraging to those just starting out. This was an important element of the nuts-and-bolts approach. Later in the morning, participants discussed typical assignments in their courses that they had brought in, with an eye to determining how they could be used as springboards for ethical discussion.

And finally, a high point of the workshop, was role playing. The entire group of participants enacted a meeting of a faculty senate at which they proposed that the university institute an ethics-across-the-curriculum program. By this time, of course, they had become adept at mounting the arguments of naysayers, as well as at responding to those arguments. The faculty senate meeting provided an effective and enjoyable end to the program.

WORKSHOP: DAYS SIX AND SEVEN

On the last two days, the participants were divided into two groups of eight to ten in separate rooms. In these groups, they critiqued each other's plans for teaching in the classroom. An instructor was present to ask questions and offer advice, but the participants did most of the talking. The seminar was another component of the nuts-and-bolts approach to help participants surmount the obstacles to introducing ethics.

On the seventh day, the participants made different presentations with new problems, homework assignments, or exam questions. This time, the emphasis was on grading. The discussion examined whether material was covered in the course, that is, whether it reflected what students had been exposed to and what they could reasonably be expected to understand. The grading was qualitative, a mode of assessment new to many of the participants. Therefore, they had to determine criteria for grading their students' responses and decide how much credit should be awarded for each point and the percentage of the course grade for ethics assignments. The purpose of focusing on these details was to make sure that grading was done.

Workshop participants were required to send Michael Davis their evaluations and their students' evaluations before they received the final portion of their stipends. Although this was a relatively small sum of money, this requirement helped bring in the evaluations.

CONCLUSION

Looking to the future, we are currently working at IIT on developing ethics courses and pilot ethics modules for graduate departments in engineering and science. The idea is to build ethics into graduate education in these fields so students will encounter ethics as an ordinary part of their graduate training. We hope that when they become faculty members themselves they will not need workshops such as these.

REFERENCES

Davis, M. 1998. Thinking Like an Engineer: Studies in the Ethics of a Profession. New York: Oxford University Press.
Rachels, J. 2002. The Elements of Moral Philosophy, 4th ed. New York: McGraw Hill.

Integrating Ethics Education at All Levels

ETHICS AS A CORE COMPETENCY

STEPHANIE J. BIRD
Science and Engineering Ethics

Engineering educators have historically believed they were only responsible for turning out technical experts. It was expected, to the extent that anyone thought about it, that engineers would pick up professional values and ethical standards and practices by observing good examples. Recently, however, the community has become aware of the need to address responsible, ethical behavior explicitly as part of engineering education. Recently revised requirements of the Accreditation Board of Engineering and Technology (ABET) state that to achieve accredited status "engineering programs must demonstrate that their graduates have an understanding of professional and ethical responsibility" (ABET, 2003). Ultimately, the goal is to ensure that engineering professionals are ethically, as well as technically, competent.

This recent attention to engineering ethics reflects, in part, the realities of engineering practice. In a survey of engineering students and practicing engineers, Robert McGinn found that 80 to 90 percent of the practicing engineers surveyed (n = 294) thought that "current engineering students [were] likely to encounter significant ethical issues in their future engineering practice." In fact, 53 to 70 percent of these practicing engineers indicated that they themselves had either "faced . . . an ethical issue in the course of [their] engineering practice" or had known a fellow engineer who had. The majority of these engineers said that they wished they "had been better prepared . . . to deal thoughtfully and effectively with [that] issue." As one might predict, more than 90 percent of the practicing engineers surveyed thought engineering students "should . . . be exposed during their formal engineering education to ethical issues of the sort that they may later encounter in their professional practice" (McGinn, 2003).

125

Knowledge of the ethical standards and values of the profession is a central and integral part of an engineer's professional life. Engineers are expected to know and behave according to professional norms; they are judged not only by their colleagues and collaborators, who naturally evaluate their worthiness as members of the community, but also by the students and trainees they teach and mentor, by funders, such as the National Science Foundation, and by society in general. In short, there is more to being an engineering professional than simply being a technical expert (McGinn, 2003). Awareness of and respect for the professional values and standards of the community is a measure of one's standing in that community.

Even though engineers are expected to act in accordance with standards and values, they are not generally taught them explicitly. Instead it is generally assumed that trainees and students will observe what senior professionals do and follow their example. Unfortunately, modeling of good behavior does not always happen, and even when it does, it may not be sufficient because learning from the behavior of another requires interpretation, which can lead to misunderstanding and confusion. Moreover, the rationale for any behavior, even exemplary behavior, is not always obvious, especially when problems are multifaceted and complex and choices must be made among competing interests and concerns. For these reasons, responsible and ethical engineering practices should be addressed explicitly. Faculty and senior members of the community are key participants in this discussion, not only because they have developed expectations regarding professional behavior, but also because they set the professional standards for the engineering community.

TEACHING ETHICS

A central question, often asked, that raises a fundamental issue is whether or not ethics can be taught. Indeed, there is a widespread assumption that "All I need to know I learned in kindergarten." But conflicts of interest, intellectual property rights, and the ownership of ideas are not commonly considered in elementary school. Fortunately, research has been done to address this question. James Rest, Muriel Bebeau, and their colleagues have shown that moral development continues at least until the end of formal education, reflecting, in part, a growing awareness and reevaluation of the individual's role in society as he or she becomes a professional (Bebeau, 1991; Rest, 1986, 1988).

A primary goal in open discussions of responsible and ethical engineering practice is to increase awareness and knowledge of professional standards. In the course of examining issues of responsible behavior, a range of acceptable practices may be identified, that is, a continuum of behaviors, from preferred through acceptable, discouraged, and even prohibited practices. In the process, the underlying assumptions of acceptable practices are revealed, as are their immediate and long-term implications. Additional goals include: (1) increasing awareness of the ethical dimensions of science and engineering; (2) providing students and trainees with experience in making and defending decisions about ethical issues; and (3) helping individuals develop strategies for addressing ethical issues and identifying resources to support decisions.

In the late 1980s and early 1990s, the National Institutes of Health (NIH) established a requirement that all pre- and postdoctoral trainees funded by NIH be given formal training in conducting and reporting research responsibly. As educational programs were developed to meet these requirements, NIH identified six characteristics of effective programs: (1) required participation, which conveys the message that responsible behavior is considered essential to the profession; (2) interactive discussions that provide ample opportunities for students to think through problems and cases; (3) the participation of many faculty members and senior professionals, demonstrating that the community as a whole values responsible behavior; (4) a focus on topics relevant to the discipline; (5) programs that begin early in the curriculum and continue throughout graduate and postgraduate education, demonstrating that standards within the community continue to evolve and that, with experience, students and trainees become more sophisticated in addressing complex problems; and (6) reinforcement of professional standards and ethical values through a variety of programs and activities, including courses, laboratory meetings, and departmental seminars. The features identified by NIH for teaching research ethics can be helpful in the development of strategies for teaching engineering ethics.

OPTIONS

There are a variety of ways to present engineering ethics, each with advantages and disadvantages. Courses provide a forum for presenting a coherent and comprehensive outline of ethical issues; courses can be marginalized, however, depending upon the level of faculty support in the department. The ethics-across-the-curriculum approach emphasizes ethical issues in all core courses, highlighting values inherent in the subject matter (Cruz and Frey, 2003; Weil, 2004). Unfortunately, faculty often feel that they do not have enough time to incorporate ethical issues and teaching modules into standard core courses. Furthermore, they often feel that they lack the expertise to raise ethical issues, although with experience, many feel more comfortable about including ethical concerns in formal classroom discussions (Cruz and Frey, 2003; Weil, 2004).

Team meetings, as well as informal discussions with advisors and mentors, can provide additional opportunities for exploring ethical issues. However, these are relatively variable, both in terms of the topics covered and the quality and depth of the discussion.

Workshops

Workshops offer opportunities for detailed considerations of ethical issues that arise in the practice of the profession. Workshops can be held in the context of professional societies or in the workplace. "Engineering practice workshops" are an adaptation of "research practice seminars," which have been held for about 10 years at the Massachusetts Institute of Technology. Participants in these seminars include junior and senior faculty, postdoctoral associates, research staff, graduate students, and undergraduates, all of whom engage in a dynamic conversation. The purpose of research practice seminars, and by extension workshops on engineering practices, is to provide a forum for faculty and senior professionals to discuss their expectations and their understanding of acceptable and unacceptable behavior in terms of specific situations and cases.

These workshops provide an opportunity for small-group mentoring, that is, for faculty and senior engineers and researchers to interact and discuss details of professional practice that are not normally covered in formal classes. Students and trainees can also express their concerns and discuss their experiences, giving the whole group a chance to identify and evaluate problems and issues and develop potential solutions. This can be informative and, ultimately, helpful for both faculty and senior professionals because the nature of the graduate and postgraduate experience may have changed significantly since "their day."

Workshops also provide an opportunity for faculty and senior professionals to discuss ethical issues with their peers; these kinds of issues are rarely discussed elsewhere until serious problems develop. Moreover, although senior professionals often agree that a particular situation is problematic and assume that the best course of action is obvious, the "obvious" answer may differ from one individual to another as a result of differing backgrounds, perspectives, and experiences. Thus in interactive workshops, both senior and junior participants can not only explore strategies for dealing with complicated issues and learn which ones have worked in the past, they can also obtain feedback in a nonthreatening, productive way.

The format of these workshops is fairly simple. The framework for examining the topic, including the primary concepts or points of contention, is presented first. This is followed by a case presentation of a real-life situation, sometimes accompanied by brief (three- to five-minute) presentations by a panel that includes a senior professional, a junior professional, a trainee, and a student, each of whom addresses an aspect of the scenario that seems significant from his or her perspective. The bulk of the workshop consists of discussions, either by the whole group or by small groups first led by a facilitator and followed by a moderated discussion by the whole group designed to harvest and critique the ideas of the small groups. In either case, the discussion usually reveals that there is more than one solution to an ethical problem—more than one acceptable solution and more than one unacceptable solution—and that the "good" solution varies with point of view.

Participants are encouraged to adopt the perspective of the "agent" rather than the "judge," that is, to identify courses of action for each character as if they were that character and to examine the implications of each choice (Whitbeck, 1998). Participants are asked to make explicit the reasons they consider a particular course of action preferable or unacceptable. The general discussion is designed to critique these ideas and the analyses of their implications both for the individual and for the profession. At the end of the workshop, participants are given "A Checklist for Ethical Decision-Making," a useful tool for evaluating and addressing ethical issues they might encounter in the future (see Appendix, p. 131).

The workshop format can be readily adapted not only for intra-institutional workshops and departmental seminars in any discipline, but also for meetings as part of the program of a professional society or as part of a team meeting or corporate workshop in the workplace.

Ethics as a Component of a Project

Another teaching strategy that emphasizes ethics as a core competency is to make it an explicit component of a project. For example, over the last three years we have incorporated ethics into a National Science Foundation-funded program, Research Experience for Undergraduates (REU), for students interested in bioengineering (Hirsch et al., 2003). The central element of the REU ethics component is that each student identify an ethical aspect or implication of his or her summer project. The students discuss their projects and associated ethical concerns with other students, include a discussion of ethical issues in their presentations at the end of the summer, and most important, select one ethical issue or implication for an in-depth written discussion.

Students in the REU program have examined a wide range of topics, from the fair allocation of credit for contribution to a project to bias in communicating research results to the humane treatment of laboratory animals in teaching and research to limitations on computer access by those who are visually impaired and people in the developing world. The REU approach can be adapted to undergraduate, master's, and doctoral thesis projects, as well as to discussions of projects in the workplace in team and group meetings.

SUMMARY

Integrating ethics at all levels of education emphasizes to students and faculty that ethics is a core competency. Experience has shown that there are several characteristics of effective teaching of ethics:

- Ethical issues must be addressed explicitly. Good role models are necessary but not sufficient for teaching ethical behavior and standards.

- Participation by senior professionals and faculty is critical because they provide expertise and experience in the discussion of professional standards and values. In addition, they clarify their expectations, thus emphasizing the importance and legitimacy of professional values and ethical standards.
- The most effective way to convey ethical practices is through interactive discussions of specific cases.
- Students (and faculty and senior professionals) learn by identifying and discussing ethical issues that arise in their own projects.
- Activities that are effective in an educational setting can be adapted for the workplace.

Explicit discussions of the responsible and ethical practice of engineering, the range of ethical issues, and the professional values and standards of the community constitute an acknowledgment of the complexity of ethical issues and the need to address them. Discussions of responsible and ethical conduct also reaffirm the responsibility of the community, individually and collectively, to address these issues as professionals.

REFERENCES

ABET (Accreditation Board for Engineering and Technology). 2003. Criteria for Accrediting Engineering Programs (2003–2004). Baltimore, Md.: Accreditation Board for Engineering and Technology.

Bebeau, M. 1991. Can ethics be taught?: a look at the evidence. Journal of the American College of Dentists 58(1): 100–115.

Bird, S.J. 1993. Teaching Ethics in Science. Pp. 228-232 in Ethics, Values, and the Promise of Science. Research Triangle Park, N.C.: Sigma Xi.

Cruz, J.A., and W.J. Frey. 2003. An effective strategy for integrating ethics across the curriculum in engineering: an ABET 2000 challenge. Science and Engineering Ethics 9: 543–568.

Hirsch, P.L., S.J. Bird, and M. Davila. 2003. Enriching the Research Experience for Undergraduates (REUs) in biomedical engineering. Pp. 283–292 in Proceedings of the 2003 American Society for Engineering Education Annual Conference and Exposition. Washington, D.C.: American Society for Engineering Education.

McGinn, R.E. 2003. "Mind the gaps": an empirical approach to engineering ethics, 1997–2001. Science and Engineering Ethics 9: 517–542.

Rest, J.R. 1986. Moral Development in Young Adults. Pp. 92–111 in Adult Cognitive Development, edited by R.A. Mines and K.S. Kitchener. New York: Praeger.

Rest, J.R. 1988. Can Ethics Be Taught in Professional Schools?: The Psychological Research. Pp. 22–26 in Ethics: Easier Said Than Done. Marina del Ray, Calif.: Joseph & Edna Josephson Institute of Ethics.

Swazey, J.P., and S.J. Bird. 1995. Teaching and learning research ethics. Professional Ethics 4: 155–178.

Velasquez, M. 1992. Business Ethics, 3rd ed. Englewood Cliffs, N.J.: Prentice Hall.

Weil, V. 1993. Teaching Ethics in Science. Pp. 243–248 in Ethics, Values, and the Promise of Science. Research Triangle Park, N.C.: Sigma Xi.

Whitbeck, C. 1998. Ethics in Engineering Practice and Research. Cambridge, U.K.: Cambridge University Press.

APPENDIX
A CHECKLIST FOR ETHICAL DECISION-MAKING[1]

STEP 1   Recognize and define the ethical issues (i.e., identify what is [are] the problem[s] and who is involved or affected).

STEP 2   Identify the key facts of the situation, as well as ambiguities or uncertainties, and what additional information is needed and why.

STEP 3   Identify the affected parties or "stakeholders" (i.e., individuals or groups who affect, or are affected by, the problem or its resolution). For example, in a case involving intentional deception in reporting research results, those affected include those who perpetrated the deception, other members of the research group, the department and university, the funder, the journal where the results were published, other researchers developing or conducting research on the findings, etc.

STEP 4   Formulate viable alternative courses of action that could be taken, and continue to check the facts.

STEP 5   Assess each alternative (i.e., its implications; whether it is in accord with the ethical standards being used, and if not, whether it can be justified on other grounds; consequences for affected parties; issues that will be left unresolved; whether it can be publicly defended on ethical grounds; the precedent that will be set; practical constraints, e.g., uncertainty regarding consequences, lack of ability, authority or resources, institutional, structural, or procedural barriers).

STEP 6   Construct desired options and persuade or negotiate with others to implement them.

STEP 7   Decide what actions should be taken and in so doing, recheck and weigh the reasoning in steps 1–6.

---

[1]From Swazey and Bird, 1995; Weil, 1993; and Velasquez, 1992.

Appendixes

Appendix A
Biographies

John F. Ahearne is director of the ethics program at Sigma Xi, the Scientific Research Society; adjunct professor of civil and environmental engineering and lecturer in public policy at Duke University; and adjunct scholar for Resources for the Future. He earned a Ph.D. in physics from Princeton University. An expert on nuclear power and nuclear weapons, Dr. Ahearne was a commissioner of the U.S. Nuclear Regulatory Commission from 1978 to 1983 (chairman from 1979 to 1981). He was deputy assistant secretary of energy in the White House Energy Office from 1977 to 1978 and deputy and principal deputy assistant secretary of defense from 1972 to 1977, working on weapons systems analysis. Dr. Ahearne is an active member of the National Academy of Engineering, American Nuclear Society, National Council on Radiation Protection and Measurements, and Society for Risk Analysis, of which he was president from 2001 to 2002. He currently chairs the National Research Council Board on Radioactive Waste Management and the American Physical Society Panel on Public Affairs. He is a fellow of the American Physical Society, the American Academy of Arts and Sciences, the Society for Risk Analysis, and the American Association for the Advancement of Science.

Braden R. Allenby is the environment, health, and safety vice president for AT&T and an adjunct professor at Columbia University. He graduated *cum laude* from Yale University in 1972, received his J.D. from the University of Virginia Law School in 1978, his M.A. in economics from the University of Virginia in 1979, his M.A. in environmental sciences from Rutgers University in 1989 and his Ph.D. in environmental sciences from Rutgers University in 1992. Dr. Allenby is coauthor or author of several engineering textbooks, including *Industrial Ecology* (Prentice-Hall, 1995), *Industrial Ecology and the Automobile* (Prentice-Hall, 1997), and *Industrial Ecology: Policy Framework and Implementation* (Prentice-Hall, 1999).

Stephanie J. Bird is coeditor of the journal *Science and Engineering Ethics*, an international publication that explores ethical issues of direct concern to scientists and engineers. The journal is widely abstracted and indexed and was recently cited by the National Academy of Sciences as a leading resource for scholarly articles on research integrity. Dr. Bird is a former special assistant to the vice president for research at the Massachusetts Institute of Technology (MIT), where she worked on the development of educational programs that address ethical issues in science and engineering, professional responsibilities, and ethical issues in research practice and science generally. She is also a laboratory-trained neuroscientist whose research interests include the ethical, legal, and social policy implications of scientific research. Dr. Bird teaches in her areas of expertise at MIT and has written numerous articles on the responsible conduct of research and mentoring and other responsibilities of science professionals. She also lectures and conducts workshops at professional societies, conferences, medical schools, and research and teaching institutions in the United States and other countries.

Charles E. (Ed) Harris earned a B.S. in biology and chemistry and a Ph.D. in philosophy from Vanderbilt University. His area of expertise is practical ethics, with a special focus on engineering ethics. In addition to a number of papers in professional journals, Dr. Harris is coauthor, with Michael S. Pritchard and Michael J. Rabins, of *Engineering Ethics: Concepts and Cases* (Wadsworth Publishing, 1995), *Practicing Engineering Ethics* (IEEE, 1997), and *Applying Moral Theories,* (Wadsworth Publishing, 1997, 3rd ed.).

Joseph R. Herkert is associate professor of multidisciplinary studies, director of the Benjamin Franklin Scholars Program (a dual-degree program in engineering and humanities/social sciences), and interim director of the Science, Technology, and Society Program, all at North Carolina State University. Dr. Herkert teaches courses in Engineering Ethics; Science, Technology and Values; Technological Catastrophes; and Technology Assessment. He received a B.S.E.E. from Southern Methodist University and a D.Sc. in engineering and policy from Washington University in St. Louis. His research interests include engineering ethics, social implications of information and communication technology, and energy/environmental policy. Dr. Herkert is editor of *Social, Ethical and Policy Implications of Engineering: Selected Readings* (Wiley/IEEE Press, 1999) and the IEEE journals, *Society on Social Implications of Technology* and *Technology* *and Society*. He is also a former president of the Society on Social Implications of Technology.

Deborah G. Johnson is the Anne Shirley Carter Olsson Professor of Applied Ethics at the University of Virginia (UVA). She joined the faculty at UVA in the fall of 2001 after three years at Georgia Institute of Technology and 20 years at Rensselaer Polytechnic Institute. Dr. Johnson is the author or editor of more than 40 published papers and four books, including a popular textbook, *Computer Ethics* (Prentice Hall, 2000), now in its third edition; the book has been translated into Spanish and Japanese. She is also coeditor of the journal, *Ethics and Information Technology*, and coeditor (with S. Rosser and M.F. Fox) of *Women, Gender, and Technology*, a series published by University of Illinois Press. Dr. Johnson recently completed a term as president of the Society for Philosophy and Technology and is currently president of a new professional society, the International Society for Ethics and Information Technology (INSEIT).

George Khushf is humanities director of the Center for Bioethics and an associate professor in the Department of Philosophy, University of South Carolina. He is also a member of the editorial boards of several journals, including *Health Care Analysis* and *Ethical Theory and Moral Practice*, and a consultant on administrative and organizational ethics for government agencies, including the South Carolina Department of Health and Environmental Control and the Department of Disabilities and Special Needs. He has published extensively on bioethics, the philosophy of medicine, and the philosophy of science and technology. Dr. Khushf is co-principle investigator of a $1.35 million grant from the National Science Foundation to study philosophical and ethical issues associated with nanotechnology. Some of his initial research will be published in a forthcoming issue of *Annals of the New York Academy of Sciences*. After receiving his B.S. in civil engineering *summa cum laude* from Texas A&M University, Dr. Khushf went on to earn an M.A. and Ph.D. in philosophy and religion.

Vivian Weil is director of the Center for the Study of Ethics in the Professions at the Illinois Institute of Technology (IIT). She received her A.B. and M.A. from the University of Chicago and her Ph.D. from the University of Illinois, Chicago. During the academic year 1990–1991, she was director of the Ethics and Values Studies Program of the National Science Foundation. Dr. Weil is a fellow of the American Association for the Advancement of Science, the Governing Board of the National Institute for Engineering Ethics, the Executive Committee of the Association for Practical and Professional Ethics, and a former member of the Committee on Computer Use in Philosophy of the American Philosophical Association. Her recent public lectures and panel presentations have dealt with ethical issues in research, intellectual property in graduate science education, conflicts of interest involving university scientists, educating scientists and engineers concerning professional responsibility, and mentoring and ethical issues in biotechnology. Dr. Weil is coeditor of *Owning Scientific and Technical Information: Value and Ethical Issues* (Rutgers University Press, 1990), editor of *Beyond Whistleblowing: Defining Engineers' Responsibilities* (CSEP, 1983), and editor of *Trying Times: Science and Responsibilities after Daubert*, produced by CSEP in collaboration with the Institute for Science, Law and Technology (ISLAT) at the IIT Chicago-Kent College of Law.

Caroline Whitbeck is Elmer G. Beamer-Hubert H. Schneider Professor in Ethics at Case Western Reserve University, where she holds appointments in the Department of Philosophy and the Department of Mechanical and Aerospace Engineering. Dr. Whitbeck earned a B.S. in mathematics from Wellesley College, an M.S. from Boston University, and a Ph.D. in philosophy from Massachusetts Institute of Technology. In the 1970s and 1980s, she initiated feminist philosophical critiques of science, delineated a feminist self-other distinction, and gave philosophical voice to ethical concerns underlying the women's health movement. In the 1980s and 1990s, she developed the analogy between ethical problems and design problems, particularly problems of engineering design and research design. She has pioneered active learning methods in the teaching of engineering ethics and the responsible conduct of research, with an emphasis on agent-centered problem solving, which has been widely adopted. Dr. Whitbeck was elected a fellow of the American Association for the Advancement of Science for her work on engineering ethics and was a Phi Beta Kappa Visiting Scholar in 1994–1995. She is the author of *Ethics in Engineering Practice and Research* (Cambridge University Press, 1998).

Paul Root Wolpe is a professor in the Department of Psychiatry at the University of Pennsylvania (Penn) and a member of the faculties of the Department of Medical Ethics and the Department of Sociology. He is also a senior fellow of the Center for Bioethics, director of the Program in Psychiatry and Ethics at the School of Medicine, senior fellow of the Leonard Davis Institute for Health Economics, and a member of the Cancer Center and Center for AIDS Research. In addition, Dr. Wolpe is the first chief of bioethics for the National Aeronautics and Space Administration (NASA); his responsibilities include ensuring that research subjects and astronauts are protected, both by NASA and by our international space partners. Dr. Wolpe is the author of numerous articles and book chapters on sociology, medicine, and bioethics and the author of *Sexuality and Gender in Society* (HarperCollins, 1996) and *In the Winter of Life* (Reconstructionist Rabbinical College, 2002). He serves on the national boards of the American Society of Bioethics and Humanities, Planned Parenthood Federation of America's National Medical Committee, and the National Embryo Donation Advisory Board of RESOLVE and is an advisor to private industry and government agencies.

Wm. A. Wulf, president of the National Academy of Engineering, is on leave from the University of Virginia (UVA), where he is AT&T Professor of Computer Science and University Professor. From 1988 to 1990, Dr. Wulf was assistant director of the National Science Foundation (again on leave from UVA). Prior to joining the faculty at UVA, he founded a software company, Tartan Laboratories, based on his research as a faculty member at Carnegie Mellon University. His research focused on computer architecture, computer security, programming languages, and the optimization of compilers. Dr. Wulf is a fellow of the American Academy of Arts and Sciences, a corresponding member of the Academia Espanola de Ingeniera, a foreign member of the Russian Academy of Sciences, and a fellow of four professional societies: Association for Computing Machinery, Institute of Electrical and Electronics Engineers, American Association for the Advancement of Science, and the Association for Women in Science. In addition, he has written more than 100 papers and technical reports and three books; he owns two U.S. Patents and has supervised more than 25 doctoral dissertations in computer science.

Appendix B
Program

Tuesday, October 14

Emerging Technologies

Moderator: Deborah G. Johnson, Anne Shirley Carter Olsson Professor of Applied Ethics, University of Virginia

9:00 a.m.   Keynote Address
Wm. A. Wulf, President
National Academy of Engineering

9:30 a.m.   The Ethical Dimensions of Earth Systems Engineering and Management
Braden R. Allenby, Environment, Health, and Safety
Vice President
AT&T

10:00 a.m.   Nano-Ethics: Framing the Issues
George Khushf, Humanities Director of the Center for Bioethics, and Associate Professor of Philosophy
University of South Carolina

10:30 a.m.   Break

141

10:45 a.m.   Neurotechnology and Brain-Computer Interfaces: Ethical and
                    Social Implications
                    Paul Root Wolpe, Senior Fellow, Center for Bioethics
                    University of Pennsylvania

11:15 a.m. E^3: Energy, Engineers, and Ethics
                    John F. Ahearne, Director, Ethics Program
                    Sigma Xi, The Scientific Research Society

12:00 p.m.   Lunch

State of the Art in Engineering Ethics

Moderator: Stephanie J. Bird, Editor, Science and Engineering Ethics

1:00 p.m.   Ethical Methodology for Case Studies in Engineering Ethics
                    Charles E. (Ed) Harris, Associate Professor,
                    Department of Philosophy
                    Texas A&M University

1:30 p.m.   Creativity and Responsibility in Engineering
                    Caroline Whitbeck, Elmer G. Beamer-Hubert H. Schneider
                    Professor in Ethics, Department of Philosophy, and
                    Department of Mechanical and Aerospace Engineering
                    Case Western Reserve University

2:00 p.m.   Microethics, Macroethics, and Professional Engineering
                    Societies
                    Joseph R. Herkert, Associate Professor of Multidisciplinary
                    Studies
                    North Carolina State University

Ethics Challenges

2:45 p.m.   Breakout Sessions on Ethics and Emerging Small- and
                    Large-Scale Technologies

4:30 p.m.   Film *Incident at Morales*
                    Developed and produced by the National Institute for
                    Engineering Ethics

5:20 p.m.   Reception

6:00 p.m.   Dinner

Wednesday, October 15

9:00 a.m.   Reports from Breakout Groups

Ethics in Engineering Education

Moderator: John F. Ahearne, Director, Ethics Program
Sigma Xi, The Scientific Research Society

10:00 a.m.   Ethics across the Curriculum
                Vivian Weil, Director, Center for the Study of Ethics in the
                Professions, and Professor of Ethics
                Illinois Institute of Technology

10:30 a.m.   Integrating Ethics Education at All Levels
                Stephanie J. Bird, Editor
                Science and Engineering Ethics

11:00 a.m.   Facilitated Discussion: Where Do We Go from Here?
                Led by John F. Ahearne, Stephanie J. Bird, Deborah G.
                Johnson, and Wm. A. Wulf

12:00 p.m.   Lunch

Appendix C
Workshop Participants

John F. Ahearne
Director, Ethics Program
Sigma Xi, The Scientific Research
    Society
Durham, North Carolina

Braden R. Allenby
Environment, Health, and Safety
    Vice President
AT&T
Bedminster, New Jersey

Sheri Alpert
Acting Director; Science,
    Technology, and Values Program
University of Notre Dame
Notre Dame, Indiana

Marc Apter
Coordinator, Ethics and Member
    Conduct Committee
Institute of Electrical and Electronics
    Engineers (IEEE)
Alexandria, Virginia

Randy Atkins
Senior Program Officer, Media and
    Public Relations
National Academy of Engineering
Washington, D.C.

Sonja Atkinson
Administrative Assistant
National Academy of Engineering
Washington, D.C.

Marc Aubertin
Lecturer, Engineering Writing
    Program
University of Southern California
Los Angeles, California

Richard Balzhiser
President Emeritus
Electric Power Research Institute Inc.
Palo Alto, California

Diana Bauer
Project Officer, National Center for
    Environmental Research
Environmental Protection Agency
Washington, D.C.

Angele Lauria Baumann
Quality Assurance Physician for
    Human Subjects Research
Brookhaven National Laboratory
Upton, New York

Daniel Berg
Institute Professor, Decision Sciences
    and Engineering Systems
Rensselaer Polytechnic Institute
Troy, New York

Rosalyn Berne
Assistant Professor, School
    of Engineering
University of Virginia
Charlottesville, Virginia

Stephanie J. Bird
Editor
Science & Engineering Ethics
Cambridge, Massachusetts

Tracy Blake
Intern
National Academy of Engineering
Washington, D.C.

Peter Bofah
Assistant Professor and Acting
    Assistant Director of CESAC,
    Department of Electrical and
    Computer Engineering
Howard University
Washington, D.C.

Eugene Brown
Program Director, Education and
    Human Resources/Division of
    Graduate Education
National Science Foundation
Arlington, Virginia

Sarah Brown
Research Associate, Board on
    Mathematical Sciences and their
    Applications
National Academy of Sciences
Washington, D.C.

Dragana Brzakovic
Staff Associate, Office of Integrative
    Activities
National Science Foundation
Arlington, Virginia

Thomas Budinger
Professor and Chair, Department
    of Bioengineering
University of California, Berkeley
Berkeley, California

Richard Case
Independent Consultant
Greenwich, Connecticut

Vivienne Chin
Senior Administrative Assistant
National Academy of Engineering
Washington, D.C.

Margaret Chu
Staff Scientist, Office of Research
    and Development
U.S. Environmental Protection
    Agency
Washington, D.C.

Eileen Collins
Director and Analyst
Science & Technology Studies
Washington, D.C.

Sarah Comley
International Observers

Richard A. Conway
Senior Corporate Fellow, retired
Union Carbide Corp.
Charleston, West Virginia

Matthew Cottle
Development Officer
The National Academies
Washington, D.C.

Michael E. Davey
Analyst in Science and Technology
Congressional Research Service
Washington, D.C.

Lance Davis
Executive Officer
National Academy of Engineering
Washington, D.C.

George Dieter
Emeritus Professor of Mechanical
    Engineering, Glenn L. Martin
    Institute Professor of Engineering
University of Maryland
College Park, Maryland

Jackson Durkee
Consulting Structural Engineer
Bethlehem, Pennsylvania

Frances Elliott
Winchester, Massachusetts

Nariman Farvardin
Dean, A. James Clark School
    of Engineering
University of Maryland
College Park, Maryland

Edith Flanigen
Independent Consultant
White Plains, New York

Paul Fleury
Dean, Faculty of Engineering
Yale University
New Haven, Connecticut

Hans Forsberg
Chairman
Aangpannefoereningens
    Forskingsstiftelse
Stockholm, Sweden

Harold Forsen
Senior Vice President, retired
Bechtel Corporation
Kirkland, Washington

Kenneth R. Foster
Professor of Bioengineering
University of Pennsylvania
Philadelphia, Pennsylvania

Victoria Friedensen
Program Officer
National Academy of Engineering
Washington, D.C.

Eli Fromm
Director of the Center for Educational
    Research
Drexel University
Philadelphia, Pennsylvania

Robert Frosch
Senior Research Fellow, Belfer
   Center for Science and
   International Affairs, John F.
   Kennedy School of Government
Harvard University
Cambridge, Massachusetts

Elsa Garmire
Sydney E. Junkins Professor of
   Engineering, Thayer School
   of Engineering
Dartmouth College
Hanover, New Hampshire

Penny Gibbs
Program Associate
National Academy of Engineering
Washington, D.C.

Paul Gilbert
Chairman Emeritus
Parsons Brinckerhoff Inc.
Seattle, Washington

Cecile Gonzalez
Senior Media/Public Relations
   Assistant
National Academy of Engineering
Washington, D.C.

Elizabeth Gordon
Ithaca, New York

William Gordon
Consulting Engineer
Ithaca, New York

Al Grant
Chair, Engineers Forum
   on Sustainability
American Society for Engineering
   Education
Washington, D.C.

Cary Gravatt
Director, Manufacturing
   Competitiveness, Technology
   Administration
U.S. Department of Commerce
Washington, D.C.

Albert C. Gray
Executive Director
National Society of Professional
   Engineers
Alexandria, Virginia

Robert J. Gustafson
President
American Society of Agricultural
   Engineers
St. Joseph, Michigan

Bill Hansmire
Vice President, Manager of
   Underground Engineering
Parsons Brinckerhoff Inc.
San Francisco, California

Charles E. (Ed) Harris
Associate Professor, Philosophy
   Department
Texas A&M University
College Station, Texas

William J. Harris, Jr.
Consultant
Arlington, Virginia

Eeva Hedefine
Graduate Research Assistant,
   Department of Spatial
   Information Science and
   Engineering
University of Maine
Orono, Maine

William (Bill) Hederman
Director, Office of Market Oversight
and Investigations
Federal Energy Regulatory
Commission (FERC)
Washington, D.C.

Otto Helweg
Dean, College of Engineering
and Architecture
North Dakota State University
Fargo, North Dakota

Laurie Henrikson
Assistant General Manager, Advanced
Technology
The Aerospace Corporation
Los Angeles, California

Joseph R. Herkert
Associate Professor of
Multidisciplinary Studies
North Carolina State University
Raleigh, North Carolina

Rachelle Hollander
Senior Science Advisor, Social
and Economic Sciences
National Science Foundation
Arlington, Virginia

Janet Hunziker
Program Officer
National Academy of Engineering
Washington, D.C.

Grace Huynh
Intern
National Academy of Engineering
Washington, D.C.

Jacqueline Isaacs
Associate Professor
Northeastern University
Boston, Massachusetts

David Japikse
Chairman and CEO
Concepts NREC
White River Junction, Vermont

Deborah G. Johnson
Anne Shirley Carter Olsson Professor
of Applied Ethics
University of Virginia
Charlottesville, Virginia

Donald Johnson
Consultant
Hertford, North Carolina

Richard Johnson
Senior Partner
Arnold & Porter
Washington, D.C.

Barbara Karn
Office of Research and Development,
National Center for
Environmental Research,
Environmental Engineering
Research Division
U.S. Environmental Protection
Agency
Washington, D.C.

Kristina Katsaros
Director, retired
National Oceanic and Atmospheric
Administration
Freeland, Washington

Michael Katsaros
Freeland, Washington

Raphael Katzen
Consulting Engineer
Bonita Springs, Florida

Selma Katzen
Bonita Springs, Florida

Maribeth Keitz
Senior Public Understanding of
    Engineering Associate
National Academy of Engineering
Washington, D.C.

George Khushf
Humanities Director of the Center for
    Bioethics, and Associate
    Professor of Philosophy
University of South Carolina
Columbia, South Carolina

Herwig Kogelnik
Adjunct Photonics Research
    Vice President
Bell Laboratories, Lucent
    Technologies
Holmdel, New Jersey

Eugene Kremer
Professor Emeritus
Kansas State University
Manhattan, Kansas

Mary Kutruff
Assistant Awards Administrator
National Academy of Engineering
Washington, D.C.

James Lammie
Director Emeritus
Parsons Brinckerhoff Inc.
New York, New York

Bryn Lander
Intern, Program on Scientific
    Freedom, Responsibility and Law
American Association for the
    Advancement of Science
Washington, D.C.

Robert Lanphier
President and CEO
AGMED Inc.
Springfield, Illinois

Darrell Laurant
Writer
Science and Spirit
Quincy, Massachusetts

Gillseung Lee
Director
Korea-U.S. Science Cooperation
    Center
Vienna, Virginia

Frances Li
Senior Staff Associate, Office
    of International Science
    and Engineering
National Science Foundation
Arlington, Virginia

Soo-siang Lim
Program Director, Division
    of Engineering Education
    and Centers
National Science Foundation
Arlington, Virginia

Frederick Ling
Earnest F. Gloyna Regents Chair
    Emeritus in Engineering
The University of Texas at Austin
New York, New York

Michael Loui
Professor of Electrical and Computer
 Engineering, Coordinated
 Science Laboratory
University of Illinois at Urbana-
 Champaign
Urbana, Illinois

Gilda Ludwig
Santa Barbara, California

Daniel Lynch
MacLean Professor of Engineering
Dartmouth College
Hanover, New Hampshire

Giulio Maier
Professor of Structural Engineering
Technical University (Politecnico)
 of Milan
Milan, Italy

Craig Marks
Retired Vice President, Technology
 and Productivity
AlliedSignal Inc.
Bloomfield Hills, MI

David Matlock
Armco Foundation Fogarty Professor,
 Department of Metallurgical and
 Materials Engineering
Colorado School of Mines
Golden, Colorado

Meg McCoy
Research Assistant
Institute of Medicine
Washington, D.C.

Simon Middelhoek
Professor Emeritus, Faculty of
 Electrical Engineering,
 Mathematics and Computer
 Science
Delft University of Technology
Delft, The Netherlands

Jim Miller
Senior Program Associate, Program
 of Dialogue on Science, Ethics,
 and Religion
American Association for the
 Advancement of Science
Washington, D.C.

Richard K. Miller
President
Franklin W. Olin College
 of Engineering
Needham, Massachusetts

Anu Mittal
Director
U.S. General Accounting Office
Washington, D.C.

Bijan Mohraz
Professor of Environmental and Civil
 Engineering
Southern Methodist University
Dallas, Texas

James Monsees
Senior Vice President, Technical
 Director and Principal
 Professional Associate
Parsons Brinckerhoff Inc.
Orange, California

Julia Moore
Senior Advisor, Office of
   International Science and
   Engineering
National Science Foundation
Arlington, Virginia

Edward Munn
Lecturer in Philosophy and
   Coordinator of the Engineering
   Ethics Program
University of South Carolina
Columbia, South Carolina

Patrick J. Natale
Executive Director
American Society of Civil Engineers
Reston, Virginia

John M. Niedzwecki
Associate Vice Chancellor for
   Engineering, Executive Associate
   Dean, Dwight Look College of
   Engineering
Texas A&M University
College Station, Texas

Stephen Norton
Assistant Director of Undergraduate
   Studies
University of Maryland
College Park, Maryland

Andrew Noyes
Reporter
Research USA
Washington, D.C.

Ahmad Nurriddin
Program Manager, Graduate Student
   Researchers Program
NASA
Washington, D.C.

Kip P. Nygren
Professor and Head, Department of
   Civil and Mechanical
   Engineering
U.S. Military Academy
West Point, New York

Samuel Oduyela
Writer
Washington Nigerian Times
Washington, D.C.

Deborah Olster
Senior Advisor, Office of Behavioral
   and Social Sciences Research
National Institutes of Health
Bethesda, Maryland

Tom O'Neil
President and Chief Operating Officer
Cleveland-Cliffs, Inc.
Cleveland, Ohio

John Orloff
Professor, Department of Electrical
   and Computer Engineering
University of Maryland
College Park, Maryland

Jamie Ostroha
Intern
National Academy of Engineering
Washington, D.C.

David Pai
President, retired
Foster Wheeler Development
   Corporation
Livingston, New Jersey

Greg Pearson
Program Officer
National Academy of Engineering
Washington, D.C.

Jeanne Petschauer
Community Involvement
    Representative
Brookhaven National Laboratory
Upton, New York

Sarah A. Pfatteicher
Assistant Dean, College
    of Engineering
University of Wisconsin, Madison
Madison, Wisconsin

Thomas M. Powers
Research Fellow, Division of TCC,
    School of Engineering and
    Applied Science
University of Virginia
Charlottesville, Virginia

Michael Prats
President
Michael Prats and Associates, Inc.
Houston, Texas

Gary Purdy
University Professor Emeritus,
    Department of Materials Science
    and Engineering
McMaster University
Hamilton, Ontario, Canada

Matthew Quint
Assistant Counsellor, Nuclear Issues
Embassy of Australia
Washington, D.C.

Henry Rachford
Senior Principal Software Designer
AdvanticaStoner
Houston, Texas

Sarah Rajala
Associate Dean for Research and
    Graduate Programs
North Carolina State University
Raleigh, North Carolina

Emery Reeves
Schriever Chair, Professor of Space
    Systems Engineering, Retired
United States Air Force Academy
Palos Verdes Estates, California

David Rejeski
Director, Foresight and Governance
    Project
Woodrow Wilson International Center
    for Scholars
Washington, D.C.

Robert Russell
Hanover, New Hampshire

Gene Russo
Freelance Writer
Takoma Park, Maryland

Roger Schmitz
Professor of Chemical Engineering
University of Notre Dame
Notre Dame, Indiana

Johanna Levelt Sengers
NIST Fellow Emeritus
National Institute of Standards
    and Technology
Gaithersburg, Maryland

Dave Sharma
Division of Technical Investigations
Federal Energy Regulatory
    Commission (FERC)
Washington, D.C.

Jennifer Slimowitz
Program Officer, Board on Math
    Sciences and Their Applications
National Research Council
Washington, D.C.

Amanda Slocum
Science and Technology Policy Intern
The National Academies
Washington, D.C.

Elva Smith
President
Applied Global Research Corp.
Washington, D.C.

Gregory Stephanopoulos
A.D. Little Professor of Chemical
    Engineering
Massachusetts Institute of Technology
Cambridge, Massachusetts

Lowndes F. Stephens
J. Rion McKissick Professor, School
    of Journalism and Mass
    Communications
University of South Carolina
Columbia, South Carolina

Gerald B. Stringfellow
Dean, College of Engineering
University of Utah
Salt Lake City, Utah

Aaron Thorp
Student
Virginia Polytechnic Institute
    and State University
Blacksburg, Virginia

Harry Tollerton
Consultant
American Society for Engineering
    Education
Washington, D.C.

Kathy Tollerton
Public Affairs Manager
American Society for Engineering
    Education
Washington, D.C.

Dat Tran
Science Policy Analyst
National Institutes of Health
Bethesda, Maryland

Janet Twomey
Program Officer, Division of Design,
    Manufacture, and Industrial
    Innovation
National Science Foundation
Arlington, Virginia

Daniel Vallero
Director, Program in Science,
    Technology, and Human Values
Duke University
Durham, North Carolina

Gregory Vassell
Retired Senior Vice President
American Electric Power Service
    Corporation
Upper Arlington, Ohio

Dolores Vestrich
Technical Director
Computer Technical Support Service
Falls Church, Virginia

Jaw-Kai Wang
Professor of Bioengineering
and Aquaculture
University of Hawaii
Honolulu, Hawaii

Christopher Washington
Associate Professor, Department
of Engineering
Norfolk State University
Norfolk, Virginia

Vivian Weil
Director, Center for the Study of
Ethics in the Professions and
Professor of Ethics
Illinois Institute of Technology
Chicago, Illinois

Jameson Wetmore
Postdoctoral Fellow
University of Virginia
Charlottesville, Virginia

Caroline Whitbeck
Elmer G. Beamer-Hubert H.
Schneider Professor in Ethics,
Department of Philosophy,
and Department of Mechanical
and Aerospace Engineering
Case Western Reserve University
Cleveland, Ohio

Paul Root Wolpe
Senior Fellow, Center for Bioethics
University of Pennsylvania
Philadelphia, Pennsylvania

Tamae Wong
Program Officer, Policy and
Government Affairs
National Research Council
Washington, D.C.

Richard Woods
Professor Emeritus of Geotechnical
Engineering, Department of Civil
and Environmental Engineering
University of Michigan
Ann Arbor, Michigan

Richard N. Wright, III
Retired Director, Building and Fire
Research Laboratory
National Institute of Standards and
Technology
Gaithersburg, Maryland

Wm. A. Wulf
President
National Academy of Engineering
Washington, D.C.

Yannis Yortsos
Senior Associate Dean, School
of Engineering
University of Southern California
Los Angeles, California

James Zwolenik
Independent Consultant
Washington, D.C.